STATISTICS
A Gentle Introduction

Frederick L. Coolidge

SAGE Publications
London • Thousand Oaks • New Delhi

© Frederick L. Coolidge 2000

First published 2000

Apart from any fair dealing for the purposes of research or private study, or criticism or review, as permitted under the Copyright, Designs and Patents Act, 1988, this publication may be reproduced, stored or transmitted in any form, or by any means, only with the prior permission in writing of the publishers, or in the case of reprographic reproduction, in accordance with the terms of licences issued by the Copyright Licensing Agency. Inquiries concerning reproduction outside those terms should be sent to the publishers.

SAGE Publications Ltd
6 Bonhill Street
London EC2A 4PU

SAGE Publications Inc
2455 Teller Road
Thousand Oaks, California 91320

SAGE Publications India Pvt Ltd
32, M-Block Market
Greater Kailash – I
New Delhi 110 048

British Library Cataloguing in Publication data

A catalogue record for this book is available from the British Library

ISBN 0 7619 5484 8
ISBN 0 7619 5485 6 (pbk)

Library of Congress catalog card number **131531**

Typeset by M Rules
Printed in Great Britain by The Cromwell Press Ltd, Trowbridge, Wiltshire

STATISTICS

I dedicated this work to the memory of my father,
Paul Lawrence Coolidge.

Contents

10 After a Significant Analysis of Variance: Multiple Comparison Tests

11 Analysis of Variance: One Factor Repeated Measures Design

12 Analysis of Variance: Two Factor Completely Randomized Design

Acknowledgments

I want to thank my students over the years for their curiosity and questions that have helped shape this book. I want to thank Brenda (Sunni) Ball for her help in previous editions. And I want to thank profusely Sharon Stewart for all of her help.

1 A Gentle Introduction

The words 'DON'T PANIC' appear on the cover of the book *A Hitchhiker's Guide to the Galaxy*. Perhaps, it may be appropriate for some of you who read this book to invoke these words as you embark on a course in statistics. Although statistics may be legendary for driving sane people mad, or less dramatically causing undue anxiety in the hearts of countless undergraduates, statistics is simply the science of the organization and conceptual understanding of groups of numbers. Statistics helps us to understand a group of numbers in a quick and efficient way. It also helps us to make conceptual sense of the numbers so that we might communicate this information about the numbers to others. As in any other scientific discipline, statistics has its own language, and it will be important for you to learn these new terms and to see how some common words have different meanings in a statistical context.

HOW MUCH MATH DO I NEED?

Can you add, subtract, multiply and divide with the help of a calculator? If you answered 'yes' (even if you are slow at the calculations), then you can handle statistics. If you answered 'no,' then you may wish to brush up on your basic math.

THE PURPOSE OF STATISTICS: UNDERSTANDING

A group of numbers in mathematics is called a set but in statistics this group is more frequently called data. Typically the numbers themselves represent scores on some test, or they might represent the number of people who show up at some event. It is the purpose of statistics to take all of these numbers or data and to present them in a more efficient way. An even more important use of statistics, contrary to some people's beliefs, is to present these data in a more comprehensible way. People who obfuscate (to bewilder or confuse) with statistics are not really representative of the typical and ethical statistician.

ANOTHER PURPOSE OF STATISTICS: MAKING AN ARGUMENT

People often use statistics to support their opinion. People concerned with reducing the incidence of lung cancer use statistics to argue that cigarettes increase the likelihood of lung cancer. Car and truck makers use statistics to show the reliability of their cars or trucks. It has been said that being a statistician is like being an honest lawyer. One can use statistics to advance an argument, support an opinion, elect a candidate, improve societal conditions, etc. Because clearly presented statistical arguments can be so powerful, say compared to someone else's simple unsupported opinion, the science of statistics can become a very important component of change.

WHAT IS A STATISTICIAN?

Although many of you would rather sit on a tack than become statisticians, imagine combining the best aspects of the careers of a curious detective, an honest attorney, and a good storyteller. Well, that is the job of a statistician (Abelson, 1995). Let us examine the critical elements of these roles in detail.

The Curious Detective: The curious detective knows a crime has been committed and examines clues or evidence at the scene of the crime. Based on this evidence, the detective develops a suspicion about a suspect who may have committed the crime. In a parallel way, a statistician develops suspicions about suspects or causative agents, like what causes Alzheimer's disease or how crowds affect helping behavior. Aspects of Alzheimer's patients' lives are quantified into numbers (data). The data become the statistician's clues or evidence, and the experimental design (how the data were collected) is the crime scene. As you already know, evidence without a crime scene is virtually useless, and equally useless are data without knowing how they were collected.

A good detective is also a skeptic. When other detectives initially share their suspicions about a suspect, the good detective typically reserves judgment until they have reviewed the evidence and observed the scene of the crime. Statisticians are similar in that they are not swayed by popular opinion. Statisticians examine the data and experimental design and develop their own hypotheses (educated guesses). However, statisticians also have the full capability of developing their own research designs and testing their own hypotheses based on the research of others.

The Honest Attorney: An honest attorney takes the facts of a case and creates a legal argument before a judge and jury. The attorney becomes an advocate for a particular position or a most likely scenario. Frequently, the facts may not form a coherent whole or the facts may have alternative explanations. A statistician is similar in that a statistician examines the data and tries to come up with a reasonable or likely explanation for why the data occurred. Ideally, statisticians are not passive people but active theoreticians, *i.e.*, scientists. Scientists are curious. They have an idea about the nature of life or reality. They wonder about relationships among variables, for example, what causes what. These hypotheses are tested through experiments. The results of the experiment are quantified (turned into numbers). Statisticians then become attorneys when they honestly determine whether they feel the data supports their original suspicions or hypotheses. If the data do support the original hypothesis, then statisticians argue their written case (study) on behalf of their hypothesis before a judge (journal editor) and jury (a critical review of their study by their peers, also known as peer review). If the statistician wins his or her case, that is, convinces the editor and peer reviewers that this hypothesis is the most likely explanation to explain the data, then the written study may be published in a scientific journal.

There are, of course, known to be unscrupulous or naive attorneys and sadly, too, unscrupulous and naive statisticians. These types either consciously or unconsciously force the facts or data to fit their hypothesis. In the worst cases they may ignore other facts not flattering to their case or even make up their data! In science, although there are a few outright cases of fraud, it is more often that we see data forced into a particular interpretation. In these cases, the role of a skeptical detective comes into play. We may ask ourselves, is this data too good to be true? Are there alternate explanations? Fortunately in science, and this is where we so strongly differ from a courtroom, our hypothesis will not be decided by one simple study. It is said that we do not 'prove' the truth or falsity of any hypothesis. It takes a series of studies, called replication, to show the usefulness of a hypothesis. A series of studies that fails to support a hypothesis will have the effect of making the hypothesis fall into disuse. There was an old psychological theory that body type (fat, skinny, muscular) was associated with specific personality traits (happy, anxious, assertive) but a vast series of studies found very little support for the original hypothesis. The hypothesis was not disproved in any absolute sense, but it fell into disuse among scientists and in their scientific journals. Ironically, this did not 'kill' the scientifically discredited body-type theory for it still lives in very popular but unscientific monthly magazines.

A Good Storyteller: Storytelling is an art. When we love or hate a book or movie, we are frequently responding to how well the story was told or how believable the story was. I once read a story about the making of steel. Now, steel-making is not high on my list (or even medium) of interesting topics but the writer unwound such an interesting and dramatic story that my attention was completely riveted. By parallel, it is not enough for a statistician to be a curious detective and an honest attorney. One must also be a good storyteller. A statistician's hypothesis may be the real one but statisticians must state their case clearly and in convincing style. Thus, a successful statistician must be able to articulate what was found in an experiment, why this finding is important and to whom, and what the experiment may mean for the future of the human race (OK, I may have exaggerated on the latter point). Articulation (good storytelling) may be one of the most critical aspects of being a good statistician and scientist. In fact, good storytelling may be one of the most inherent interests in the history of humankind.

There are also many examples of good storytelling throughout science. The origin of the universe makes a very fascinating story. Currently, there are at least two somewhat rival theories (theories are bigger and grander than hypotheses but essentially theories at their hearts are no more than educated guesses): In one story, a supreme being created the universe from nothingness in six days, and in the other story, the Big Bang, the universe started as a small egg which exploded to create the universe. Notice that each theory has a fascinating story associated with it. One problem with both stories is that we are trying to explain how something came from nothing, which is a logical contradiction. It has been suggested that the object of a myth is to provide some logical explanation for overcoming a contradiction; however, this may be an impossible task when the contradiction is real. Still, both of these theories remain very popular among lay people and scientists alike, in part, because they make interesting, provocative, and fascinating stories.

Science is not replete with uninteresting stories, or it shouldn't be even if it is. Even the reproductive behavior of planaria or the making of steel should be told in an interesting and convincing manner, *i.e.*, it should make a good story.

LIBERAL AND CONSERVATIVE STATISTICIANS

As we proceed with this course in statistics, you may come to realize that there is considerable leeway in the way data are investigated (experimental methodology and design), and in the way data are reported (statistical analyses). There appear to be two camps, the liberals and the conservatives. Just as in politics, neither position is entirely correct in the realm of statistics. Both philosophies have their advantages and disadvantages.

Scientists as a whole are generally conservative, and since statisticians are scientists, they too are conservative, as a general rule. Conservative statisticians stick with the tried and true. They prefer conventional rules and regulations. They design experiments the same historical way, and they interpret and report their interpretations in the same acceptable fashion. This position is not as stodgy as it may first appear. The conservative position has the advantage of being more readily accepted by the scientific community (including journal editors and peer reviewers). If one sticks to the rules and accepted statistical conventions, and one argues successfully according to these same rules and conventions, then there is typically a much greater likelihood that the findings will be accepted, published, and receive attention from the scientific community. Conservative statisticians are very careful in their interpretation of their data. They guard against chance playing a role in their findings by rejecting any findings or treatment effects that are small in nature. Their findings must be very clear or their treatment effect (like from a new drug) must be very large in order for them to conclude that their data are real and, consequently, there is a very low probability that pure chance could account for their findings.

The disadvantages of the conservative statistical position is that new investigative research methods, creative statistical analyses, and radical conclusions are avoided. For example, in the real world, sometimes new drug treatments are somewhat effective but not on everyone, or the new drug treatment may work on nearly everyone but the improvement is modest or marginal. By always guarding so strongly against chance, conservative statisticians frequently end up in the position of 'throwing out the baby with the bath water.' They may end up concluding that their findings are simply due to chance when in reality something is actually happening in the data that is not due to chance.

Liberal statisticians are in a freer position. They may apply exciting new methods to investigate a hypothesis and apply new methods of statistically analyzing their data. Liberal statisticians are not afraid to flaunt the accepted scientific statistical conventions. The drawback to this position is that scientists, as a whole, are like people, in general. Many of us initially tend to fear new ways of doing things. Thus, liberal statisticians may have difficulty getting their results published in standard scientific journals. They will often be criticized on the sole grounds of investigating something a different way and not for their actual results or conclusions. In addition, there are other real dangers from being a statistical liberal. Inherent in this position is that they are more willing than conservative statisticians to view small improvements as real treatment effects. In this way, liberal statisticians may be more likely to discover a new and effective treatment. However, the danger is that they are more likely to call a chance finding a real finding. And if scientists are too hasty or too readily jump to conclusions the consequences of their actions can be deadly or even worse. What can be worse than deadly? Well, consider a tranquilizer called Thalidomide in the 1960s. Although there were no consequences for men, over 10,000 babies were born to women who took Thalidomide during pregnancy, and the babies were born alive but without hands or feet. Thus, scientific liberalism has its advantages (new, innovative, creative) and its disadvantages (it may be perceived as scary, flashy, bizarre, or have results that are deadly or even worse).

Neither position is a completely comfortable one for statisticians. This book will teach you the conservative rules and conventions. It will also encourage you to think of alternative ways of exploring your data. But remember, as in life no position is a position, and any position involves consequences.

DESCRIPTIVE AND INFERENTIAL STATISTICS

The most crucial aspect of applying statistics consists of analyzing the data in such a way as to obtain a more efficient and comprehensive summary of the overall results. To achieve these goals, statistics is divided into two areas, descriptive and inferential statistics. At the outset, do not worry about the distinction between them too much: The areas they cover overlap, and descriptive statistics may be viewed as building blocks for the more complicated inferential statistics. **Descriptive Statistics** is the historically older of these two areas. It involves measuring data using graphs, tables, and basic descriptions of numbers such as averages or means. These universally accepted descriptions of numbers are called

parameters, and the most popular and important of the parameters are the mean and standard deviation.

Inferential Statistics is a relatively newer area which involves making guesses (inferences) about a large group of data (called the population) from a smaller group of data (called the sample). Typically the sample data is randomly drawn from the population or larger group of data. The concept of random sampling means that every datum or person in the population has an equal chance of being chosen for the limited size sample.

EXPERIMENTS ARE DESIGNED TO TEST THEORIES AND HYPOTHESES

A theory can be considered a group of general propositions which attempts to explain some phenomenon. Typically theories are also grand, that is, they tend to account for something major or important, like how children acquire language, how the universe began, or the nature of reality. A good theory should also provoke people into thinking. A good theory should also be able to generate testable propositions. These propositions are actually guesses about the ways things should be if the theory is correct. Hypotheses are specifically stated propositions created from and consistent with the theory which are then tested through experimental research. Theories whose specific hypotheses are frequently supported by research tend to be regarded as useful. Theories whose hypotheses fail to receive support tend to be ignored. It is important to remember that theories and hypotheses are never really proven or disproven in any absolute sense. They are either supported or fail to receive support from research. In the real world it is also frequently the case that some theories and hypotheses receive mixed results. Sometimes research findings support the propositions and other experiments fail to support them. Scientists and statisticians tend to have critical attitudes about theories and hypotheses because the repercussions in science for a bad theory can be deadly or even worse. Thus, statisticians like well-designed and well-thought out research, such that the findings and conclusions are clear, compelling, and unambiguous.

ODDBALL THEORIES

Science is not simply a collection of rigid rules. Scientific knowledge does not always advance smoothly but typically moves along in sputters, stops and starts, dead ends, and controversy. Carl Sagan, the late astronomer, said that science requires the mating of two contradictions: a willingness to think about new, unique, strange, or even bizarre explanations coupled with rigorous skepticism and hard evidence. If scientists only published exactly what they thought they would find, new discoveries would be exceedingly rare. Thus, scientists must be willing to take risks and dare to be wrong. A physicist, Wolfgang Pauli, wrote that being 'not even wrong' is even worse than being wrong because that would imply that one's theory is not even worth contradiction or dispute. Recently the Nobel Prize for Medicine was awarded to an American neurologist, Stanley Prusiner, who proposed that infectious particles called prions do not contain any genes or genetic material yet they reproduce and probably cause 'mad-cow disease' (a dementia-like disease caused by eating infected beef). For years Prusiner's ideas were considered revolutionary or even heretical; however, most research now supports his 22 years of experimental work. His 'oddball' theory was provocative, generated many testable hypotheses, and was supported by his own rigorous testing and skepticism, and scientific knowledge has benefited greatly from his theory. However, remember there is no shortage of oddballish people creating oddball theories. A good theory is not only interesting and provocative but must also be supported by testable hypotheses and solid evidence.

BAD SCIENCE

In my Sunday newspaper, there is a 'Fitness Column,' and a recent question concerned adhesive strips that go across the bridge of the nose. These are the same strips that seem to be so popular among professional athletes. 'Do they work?' asks the letter writer. The magazine's expert says they reduce air flow resistance 'as much as 30%.' However, because the 'expert' says it's true does that make it true? Of course not. Nevertheless, we are bombarded daily by a plethora of statistics, like one in four college-age women have been raped, marriage reduces drug abuse, classical music boosts IQ, and left-handers die younger. We even pass down 'facts' as truths such as Eskimos have 30 words for snow which appears not to be the case. Eskimos apparently have just as many descriptions of snow as other people. Since most of us have been raised to tell the truth, we typically do not challenge the facts or statistics fed to us daily. However, let us be realistic. Many studies are poorly designed, statistics can be misused or manipulated, and many people in our society are not interested in the truth but they are interested in money and power (that brings more money). So let us return to the 'nose strips.'

Why did someone ultimately create nose strips? Probably to make money. Helping people breathe right was perhaps, if we are very lucky, secondary. In fact, it is easy to imagine that many products on the market are not even remotely designed to do what they claim. Their sole purpose is to make money. However, let us give the nose strip creator the benefit of the doubt. Why should we use them? Well, the nose strip propaganda says they have scientific evidence that they work, and they present the following 'facts': Breathing takes 10% of our total energy, and their nose strips reduce air resistance by up to 30%. With regards to this first fact, I think it would be fair to know how total energy was measured. Isn't it rather difficult to come up with a single measure of energy expended? And even if we could come up with an acceptable measure, did they measure this 30% expenditure when people were exercising or were they at rest? If they were exercising (because I assume that's when we'd want to use nose strips), how did they measure total energy expended when the person was running around? Did they remotely or telemetrically send the energy expenditure information? I am not sure that this is even possible. It's easier for me to imagine the person at rest when they were measured for energy expenditure. Therefore, at rest measures may not be appropriate to someone who is actually running about. Furthermore, who was the person tested? What was that person's age? Will we be able to generalize to the average person? Was the participant paid? Some people will say anything for money. An even worse scenario is that some people will even deceive themselves about how well something works if they are paid enough money (curiously, Festinger, a social psychologist, found that some people will deceive themselves for too little money).

The same criticism appears to hold for the second 'fact.' How did they ever measure this 30% air resistance reduction? Was that 30% over a period of time? Was it a mean score for a large number of participants? Was it the median score? Who did the testing? Was it an independent testing group or did the nose strip people conduct the research?

A second reason the nose strip people say that we should buy their product is that Jerry Rice, the great football player, wears them. Now, we are tempted to use them because of an expert opinion and the prestige associated with that expert. But does that mean they really work? Of course not, but we rarely challenge the opinion of a supposed expert, and many factors are operating here, like the power of identification. Some people think that if Jerry uses them, then I'll be as great as Jerry if I use them. We can see as we closely examine our motivations that few of them are based on rational or logical reasoning. They are based, however, on unconscious and powerful forces that cause us to believe, trust, imitate, and follow others, particularly people whom we perceive as having more power or status than we have. This willingness to believe on sheer faith that a product or technique works has been called the **placebo effect**.

In summary, when we closely examine many studies, we find that they sorely lack any unbiased or scientific validity. And if we blindly accept facts, figures, cures, and snake-oil, we are putting ourselves in a very dangerous position. We may make useless changes in our lives or even dangerous changes. We may also unnecessarily subject ourselves to needless stress and worry.

Now, let us examine some essential questions that we should ask of any survey or study.

Who was surveyed or studied? Remember, most of the time we hope that we can use the product or would benefit from the treatment or technique. A majority of studies still use only men. Would we be able to generalize to women if only men are studied? Would we be able to generalize to children if only adults are studied? Would we be able to generalize to people if only animals are studied? Will the product be safe for people if we do not use animals in our study? If our moral values prevent us from using animals in research, what other methods are available to ensure the safety of a product?

For a sample to be representative of the larger group (the population) it should be randomly chosen. Did the study in question use random sampling? Random sampling implies that everyone in the population had an equal chance to be in the sample. If we telephone a random sample of voters in our county, is that a random sample? No, because not everyone has a telephone and not everyone has an equal chance to answer the phone if we call at 11 A.M.

Why did the people participate in the study? Did the people volunteer or were they paid? If they were paid, and paid a lot, they might skew the results in favor of the experimenter's hypothesis. Even if they were not paid, participants have been known to unconsciously bias an experiment in the experimenter's favor. Why? Because sometimes we just like people, we'd like to help them out, or we'd be embarrassed if they failed in front of us. We may not wish to share their humiliation. Or it is easy to imagine some participants who would like to see the experimenter fail because they enjoy other people's suffering and humiliation. Ultimately, it is in the experiment's best interest if the experimenter's hypothesis is kept a secret from the participants. Furthermore, the experimenter should not know who received the experimental treatment, so that the experimenter isn't biased when he or she are assessing the results of the experimental treatment or product being tested. If the experimenter doesn't know who was in what group (but someone important does) and the participants do not know what is being tested, the study is said to be a **double-blind** experiment.

Was there a control group and did the control group receive a placebo? Even if the participants were not paid and were blind to the experimenter's hypothesis, we sometimes wish to change so badly that we change even if the product or treatment in question does not really work. This is known as the **placebo effect.** Placebo effects are well-known throughout the scientific world, and they are real. In other words, people really do get better when they are given fake substances or sugar pills. Some people really do get better when we shake rattles in front of their faces or wave our fingers back and forth, or light incense and chant. Our current scientific method calls for a control group if we are testing some drug, product, or new type of psychotherapy and, furthermore, the new drug or technique should be superior to the improvement we frequently see in the control group that receives only the placebo. Thus, the nose strips should have been tested against a group of participants who received some kind of placebo (like a similar piece of tape on their noses) and both groups should have received the exact same instructions. Interestingly, the word 'placebo' means 'I shall believe' in Latin. It is the first word in evening prayers said for the dead. In about the 12th century these prayers became know as 'placebos.' People began to hire professional mourners who would say and sing the 'placebos' and relatively quickly the word came to have an unflattering and pejorative meaning which has persisted to modern times (see Brown, 1998, for a provocative discussion of the placebo effect).

How many people participated in the study? There is no firm set of rules of how many people should be studied. However, the number of people participating can have a profound effect on our conclusions. In surveys, for example, the more the merrier. The larger the number of people sampled, the more likely that the sample will be representative of the population. Thus, to some extent, the size of our sample will be determined by the size of the population to which we hope to generalize. For example, if we are interested in a study of tax paying U.S. citizens, then a sample size of 500 might still be considered a small sample if there are 100 million taxpayers. On the other hand, 500 might be considered an unnecessarily large sample if we are attempting to study the population of Colorado Buddhists.

In studies where we employ the scientific method using an experimental group and a control group, large sample sizes may actually be harmful to the truth. This occurs because as we increase the number of participants in each group, there is a peculiar statistical artifact that increases the chances we say the product works when it really does not. Increasing the sample sizes unnecessarily is called an **abuse of power**. Power in this context is defined as the ability to detect a real difference in the two groups due to the treatment. Rarely, if ever, will the experimental group be equal to the control group even if the treatment does not work! This occurs because of simple chance differences between the groups. However, if an experimenter increases the sample sizes (to 100 or more in each group) then chance differences will be seen as real differences in most statistical tests. The way to avoid an abuse of power is to do a statistical test known as **power analysis** which can help the experimenter choose an appropriate sample size.

There is also the opposite situation where a real difference between two groups might not be detected because of insufficient power or too few participants are used in each group. In this situation, the treatment may actually work but we might conclude that it does not because we haven't used enough people in each group. Again a power analysis should help in this situation. Most experimenters use a simple rule of thumb: There should be at least 10 participants in each group or the overall experiment should include at least 30 participants.

How were the questions worded to the participants in the study? Remember the 'fact' that one in four college women has reported being raped? How did the experimenter actually define the word 'rape?' In this case, the experimenter asked 6,159 college students, 'Have you ever had sexual intercourse when you didn't want to because a man gave you alcohol or drugs?' We might argue that this definition of 'rape' is too broad. It is possible that because of a woman's religious beliefs or family values, she may be ambivalent about sex (some part of her didn't want to under ANY circumstances outside of marriage). The experimenter in this situation defended her broad definition of rape because she said that rape has so many different meanings. She took the interesting position that if a student's report met the experimenter's definition of rape then the student was raped and 'whether they realize it or not is irrelevant.'

How about the 'fact' stated on milk cartons that over 1,000,000 children have been abducted? Here, the debatable issue is the definition of 'abducted.' Some missing children's organization defined abducted as any child not living with their legal custodian parent. Thus, it might be argued that some of the 'abducted' children are living with one of their biological parents, however, just not the parent who has legal custody. While this may still be a serious matter, it might not reach the same level of importance as a child who has been taken by someone unknown to the family and the whereabouts of the child are a complete mystery. If we consult police and FBI records, we would find that there are less than 1,000 children who meet the latter criterion. The problem with this broad definition of abduction is that if too many children meet the definition of abducted then we will not have the time and resources to hunt for the truly missing children.

The sexual abuse of children is certainly an important issue and highly worthy of our attention. Recently, it was stated that 25% of all males have reported the tendency to sexually abuse children. This statistic might be highly alarming except that the men surveyed were asked, 'If you knew you would never get caught or you knew that no one would ever find out, have you ever had the fantasy of sexually fondling a girl under the age of 14?' You might still consider the 25% of men who said yes to this question to be highly despicable, perverts, or sick, yet it could be argued that the question has the word 'fantasy' in it. By definition a fantasy is something not occurring in reality. It seems unfair to ask someone about a fantasy, and then assume that it reflects a true tendency in reality. Furthermore, perhaps the men were recalling an old fantasy, one they had when they were under the age of 14. What would have happened if we had asked the question like this: 'Would you ever consider having sexual intercourse with a child if you knew you could receive the death penalty, and there was a very high probability that you could get caught, and your mother would be the first person to know about it, and she would then commit suicide because you broke her heart?'

A related issue in measurement is how clearly the experimenter has defined the problem he or she trying to study or solve. Have you heard the 'fact' that one in three children goes to bed hungry? How clearly did the experimenter define what he or she meant by hungry? Since I go to bed late every night, I nearly always go to bed hungry. How did the experimenter define this word 'hungry?' Did he or she ask every child in the world if they were hungry before they went to bed? I doubt it. And what if the experimenter defined hungry by asking the child, 'Would you like anything else to eat, like a cookie or candy, before you go to sleep?' It is easy to imagine that one definition of hungry is whether a child asks for more to eat, but then many overweight children would ask for more to eat. Perhaps, you can perceive the monumental task involved in clearly defining any variable, even words as seemingly simple as hungry.

Was causation assumed from a correlational study? Probably the single most pervasive abuse of the understanding of statistics occurs when people infer causation from a correlational study. A correlational study is one in which two factors or variables are simply measured together in a setting. In correlational studies, there is no random assignment of the participants to two groups, there is no treatment applied, and there is no control group. In the correlational study if the two measured variables do appear to be related to each other many people unfortunately assume that one of these variables causes the other. While it is **remotely** possible that there is a causative relationship, a correlational study should never be interpreted as evidence for causation. For example, a correlational study of marriage and drug use found that drug use declined after marriage. The headlines in the newspapers read, 'Responsibilities of marriage and family often curb drug use.' One of the problems in this interpretation is that the study was a correlative one, and we must always guard against the inference of causation. It would be equally wrong, but no less absurd, to assume that drug use curbs marriage. A second problem with correlational studies is that many other factors or variables are also operating and may contribute as a cause. If a scientist does not measure these other variables, then the scientist cannot assume that it is only a single variable like marriage curbing drug use. In fact, an overwhelming majority of behaviors have multiple and complex causes and to assume that a single variable like alcohol use causes crime is not only statistically wrong but wrong in theory as well.

Who paid for the study? In an ideal world, the investigators in a study should not have any financial ties or interest to the outcome of a study. However, a few years ago a study reported that balding men were more likely to have heart attacks than nonbalding men. Who financed the research? A drug company that manufactures a popular hair-growing product. While this financial interest in the study does not automatically invalidate the study, it does tend to cast suspicion on the outcome of the study or the motives of the investigators.

More recently, a controversy began when a study linked a high blood pressure medicine to an increased risk of heart attacks. It was noted that 70 scientists, doctors, or researchers came to the defense of this particular blood pressure medicine in journal articles, reports, or medical publication commentaries. When the 70 were questioned about their defense, 67 of the 70 admitted they had financial relationships with the manufacturer of the drug which included travel grants, public speaking fees, research and educational grants, and consulting fees and contracts.

It has been suggested that scientists have been naive about the extent to which financial relationships may affect research. Again, in an ideal world there should be no financial complications between the investigators and the outcomes of their studies. However, short of the ideal, perhaps scientists should report any appearances of a conflict of interest at the outset of their study, either to the journal editor to which the study is submitted or in a footnote in the published version of the study. In this way, the readers of the articles can determine for themselves to what extent they feel the studies may be biased.

Was the study published in a peer-reviewed journal? Nearly all scientific journals use a peer-review procedure where any article submitted to an editor is sent out to other scientists for their opinion on whether the study should be accepted for publication. Some journals reject as many as 90% of all the articles submitted while some of the 'easier' journals may reject only 50%. Even if an article is accepted for publication (a process that takes 3–6 months or longer), it is rarely accepted without any changes. Typically, a journal editor will summarize the criticisms of the peer-reviewers and forward them to the authors. The peer-reviewers may request additional information about the literature surrounding the issue, more information about the participants, more information about the tests employed or the procedures, additional statistical analyses, or modifications of the conclusions or implications of the study. The first study that ever produced evidence for cold fusion, a process where nearly limitless energy can be obtained from a special kind of water, was not published in a peer-reviewed journal. It was presented by a teleconference! Thus, the specific procedures were not revealed in detail, and the study was not subject to any acceptable form of peer-review. Is it any wonder that the issue of whether those scientists actually created cold fusion is in question?

What about books? Do they undergo peer-review? Typically not. It does vary from publisher to publisher but there are some publishing companies where anything you write can be published in a book (for a fee). Many scientists will present the results of their research at national or international conventions. Are convention presentations peer-reviewed? Most convention presentations do have some form of peer-review but the standards for convention presentations are much more lax than for journal articles and the acceptance rates can vary from 50% to 100%. Thus, published journal articles meet the highest standards of scientific scrutiny, convention presentations are a distant second, and some books will meet absolutely no scientific standards. And we must consider paid advertisements in newspapers and magazines as well as infomercials on television as the least likely venues for truth or justice. Let the reader or viewer beware!

ON MAKING SAMPLES REPRESENTATIVE OF THE POPULATION

In order to ensure that the hypotheses we make from the sample about the population are good ones, there are two requirements: First, the sample should be randomly drawn from the population, and second, the sample should be relatively large. Remember, however, these two guidelines do not guarantee that a sample will be representative of the population. There will always exist the possibility that a sample, although large and randomly drawn, may still lack some important characteristics that are present in the population. Perhaps this is why single experiments can rarely ever prove a hypothesis or a theory. It takes repeated experiments by different experimenters before scientists begin to accept that a particular hypothesis may or may not be useful in explaining some phenomenon.

EXPERIMENTAL DESIGN AND STATISTICAL ANALYSIS AS CONTROLS

In order to obtain useful and meaningful information from both descriptive and inferential statistics, it is necessary for the data to be collected on the basis of an explicit and appropriate design called the **experimental design**. Essentially, an experimental design is a blueprint for how the data collection will be conducted. At the very beginning of this data collection or experiment there are two forms of control, or forms of power that we have at our disposal, **experimental control** and **statistical control**. Experimental control concerns the design of the experiment. For example, how many people will be used, how many groups will they be divided into, what kind of people will be used, how will they be tested or measured, etc. The second form of control or power is statistical control and this is employed after the data collection or the experiment is complete. It is important to remember that although statistical techniques are powerful tools for describing data, experimental control is the more important of the two. One reason for this is that statistics can become meaningless and will communicate no knowledge whatsoever unless we adhere to some basic experimental principles.

For example, a recent TV poll asked viewers to call two different phone numbers. The first phone number was to be called if the viewer thought that a particular university football team should be ranked as the best team in the country, and the second number was to be called if the viewer thought that the same team should not be top-ranked. This plan for the experiment is called the experimental design. The statistical analysis consisted of comparing the number of calls for the two phones and seeing which was higher. In this case the experimental design is so poor that the statistical analysis is rendered meaningless. One of the many problems with this design is that the sample of TV viewers was not random. Therefore, any conclusions that the results reflect popular opinion will be false. Another problem is that a viewer could call in more than once, and thus influence the poll. The statistical analysis cannot take into account this problem or variable. These uncontrolled problem variables in experimental designs are called **confounding variables**. In summary, the major experimental design problems with this poll are that some people were allowed to participate more than once, and the sample of viewers was not random. Thus, one aspect of this study was that it resulted in futile and meaningless data. Considerable time and money was spent (each phone call cost the viewer fifty cents), yet nothing can be concluded other than one number was higher than another number. It cannot be concluded that more people thought XYZ university was the number one team because the network cannot be certain that every phone call came from a different individual every time. In order to be ethically fair in the conclusion, the best the network might say is that the YES line received more phone calls than the NO line. However, if this conclusion gave the implication that XYZ university was the choice of most people, then even this limited conclusion is absolutely unethical. What could the network conclude after this tremendous expenditure of energy? The network could announce that the number 119,351 (phone calls on line one) is higher than 112,729 (phone calls on line two).

Now you might view these results as ridiculous, and that is exactly how they should be viewed. The experimental design in this example was so poor that no statistic could make the data interpretable. Thus, the example shows how you must think about designing your experiment in terms of your statistical analysis. Although statistical control is powerful, it cannot save or make interpretable a shoddy, poorly designed experiment.

THE LANGUAGE OF STATISTICS

Learning about statistics is much like learning a new language. Many of the terms used in statistics may be completely new to you, like **quartile**. Also, some words you already know will be used in new combinations, like **standard deviation**. Finally, some words will have a new meaning when used in the context of statistics. For example, the word **significance** has a different meaning in a statistical context. When used outside of statistics, significance is used as a value judgment. If something is said to be significant, then we usually mean that it is important or of consequence. The opposite of significant is usually insignificant. However, in statistics, the word significance has a different meaning. Significance refers to an effect that has occurred that is not likely due to chance. In statistics, the opposite of significance is nonsignificant, and this means that an effect is likely due to chance. It will be important for you to remember that the word insignificant is a value judgment, and it typically has no place in your statistical vocabulary.

ON CONDUCTING SCIENTIFIC EXPERIMENTS

Scientific experiments are generally performed to test some hypothesis. Remember, a theory is some grand idea about the way nature (at any level) seems to work. A theory generally should make some predictions in the forms of hypotheses. The hypotheses are then tested through scientific experiments. The typical experiment is a two group experimental design where there is an experimental group and a control group. Most typically, the experimental group gets or receives some special treatment. The control group usually does not get this treatment. Then, the two groups are measured in some way. These two levels of treatment are aspects of the **independent variable**. The independent variable is the variable that the experimenter manipulates. The experimenter wishes to see whether the independent variable affects the **dependent variable**. The test that the two groups are measured on is called the dependent variable. Sometimes it is very important that the participants in each group do not know whether they are receiving some special treatment or not. If this experiment was to determine whether Vitamin C affects colds, then the control group participants should also be given the exact same treatment as the experimental group with the exception of actually receiving Vitamin C. Typically, control groups receive a **placebo**, which is a pill indistinguishable from the pills which contain Vitamin C that the experimental groups receive. If the participants are not aware of who is actually receiving Vitamin C, then they are said to be **blind** to the treatment effects. It is frequently important that the experimenter also not know who is receiving Vitamin C or the placebo until after the test has been given to both groups and scored. This way the experimenter cannot inadvertently (or on purpose) affect the outcome of the experiment.

THE DEPENDENT VARIABLE AND MEASUREMENT

Remember that the test or whatever we use to measure the participants is called the dependent variable. In the Vitamin C experiments, there are a number of different measures which we might use as dependent variables. Some experimenters have focused on the number of new colds in a one year period. We could also measure the number of days sick with a cold. The choice of the dependent variable could be very important with regard to the conclusions of our study. For example, some studies have shown Vitamin C to be ineffective in preventing colds; however, there is some evidence that Vitamin C may reduce the severity or the duration of a cold.

OPERATIONAL DEFINITIONS

If we are going to count the number of colds each participant gets in a one year period, then each participant would somehow then be checked for how many colds he/she got. Is there some subjectivity in how you decide whether a person has a cold or not? Of course there is! In order to minimize this subjectivity, perhaps the same person (like a doctor of medicine) would rate all of the participants to determine the presence or absence of a cold. However, doctors might vary among themselves about whether a person actually has a cold or not. Perhaps the diagnoses might vary as a function of what each doctor defines as a cold. For example, one doctor's definition of a cold might be whether his or her patient claims to have a cold. Another doctor's definition might require the presence of congested sinuses and elevated white blood cells. There is usually no single correct definition; however, whatever definition you do choose should be stated clearly according to some criteria. When you list the criteria clearly in your study, then those criteria are said to be your **operational definition** of a cold. Therefore, it can be assumed that every participant in your study met these criteria.

MEASUREMENT ERROR

No dependent variable will ever be perfectly measured for each participant. The variation in the dependent variable that is not due to the independent variable is called **error**. The individual participants contribute the most error to the experiment and to the dependent variable. Other sources of error may come from subtle variations in the testing condition like time of day, temperature, extraneous noise levels, humidity, equipment problems, and a plethora of other minor factors. In our Vitamin C study it is possible that some participants never get colds anyway, and some may frequently get sick. The experimental error associated with the participants is hopefully balanced out when we randomly assign the participants to the two conditions. After all, we could not, for example, assign all our friends to the experimental group and all others to the control group. Random assignment of the participants to the groups helps to balance out the effects of the participants as a source of error. Using a large number of participants in each group also helps to balance out this source of error.

When you hear about or read about experiments or surveys in any form of media, remember to look at the dependent variable closely. Sometimes the entire interpretation of an experiment may hinge upon an adequate dependent variable. How about the results of a study that some city, say 'city x' is the best place to live. One critically important variable in this study is how they measure 'best city?' Is it in terms of the scenery? What happens if you like ocean views or hate ocean views? Typically, these studies try to be 'scientific' so their judges rate the cities on a number of different variables. However, many of these questions may relate to economic growth. Therefore, if a city is not growing rapidly it may not fare well in the rating system. Would you consider a fast growing city the best place in which to live? It becomes obvious there are many intangibles in developing a survey to rate the best place in which to live, and any survey which claims to be able to rate the best city is highly questionable. Be sure to examine the operational definitions of critical variables in experimental studies. In this latter example, it would be interesting to see the specific operational definition of 'best city.' If the authors do not provide an operational definition then the study is virtually uninterpretable.

MEASUREMENT SCALES: THE DIFFERENCE BETWEEN CONTINUOUS AND DISCRETE VARIABLES

A variable is typically anything that can change in value, and a variable usually takes on some numeric value. Statisticians most commonly speak of **continuous and discrete variables**. A continuous variable can be measured along a line scale which varies from a small number to a large number. For example, a continuous variable would be the time in seconds in an experiment for a participant to complete a task. A discrete variable, for statisticians, typically means that the values are unique and separate from one another. Speculations are not made between the values. For example, gender could be considered a discrete variable. If the category 'male' is assigned a value of '0' and 'female' is assigned a value of '1,' then interpretations will not be appropriate for values between '0' and '1.'

It should also be noted, when there are only two values of a discrete variable, it is referred to as a **dichotomous variable**.

TYPES OF MEASUREMENT SCALES

There are various kinds of measurement scales. In any statistical analysis, the type of scale must be identified first, so the appropriate statistical test can be chosen. Some measurement scales supply minimal information about their respective participants, and thus, the statistical analysis may be limited. Other types of scales supply a great deal of information and, consequently, the statistical possibilities are enhanced.

Nominal Scales. Nominal scales assign people or objects to qualitatively different categories. Nominal scales are also referred to as categorical scales or qualitative scales. One example is the assignment of people to one of the two categories of gender. Thus, when measured on a nominal scale, all of the people or objects in a category are the same on some particular value. Notice also that the people in the category are all considered equal with respect to that value. For example, all the males who fit the male category are considered equal with respect to the category. Thus, membership in a category does not imply magnitude. Some males in the category are not considered more 'male' than other males in the same category. The frequency of each category can be analyzed, however. For example, it might be noted that 37 males and 44 females participated in a study. Another example of a nominal scale would be a survey question that required the answer 'yes' or 'no,' or 'yes,' 'no,' or 'undecided.'

It is also important to note there are no intermittent values possible on a nominal scale. This means if, for statistical purposes, a value of '0' is assigned to the male category and '1' is assigned to the female category, there are no values allowed or assumed between the '0' and '1.' It is, of course, physiologically possible to have a person who does have a mixed gender identity. It is also psychologically possible to have a mixed gender identity. However, it is not possible to have intermittent values if gender is measured on a nominal scale.

Ordinal Scales. Ordinal scales involve ranking people on some variable. The person or object that has the highest value on the variable is ranked 'number 1,' etc. The ordinal scale, therefore, requires classification (how much of the value does an individual have), and ranking (where the individual stands relative to all other members of the group).

The ordinal scale has one major limitation, and that is, the differences between rankings may appear equal when in reality it is known that they are not. For example, if we rank athletes after a race, the difference in times between the first and second place athletes may be huge, while the difference in times between second and third place may be very small. Nevertheless, with an ordinal scale the appearance is given that the difference between the first, second, and third rankings are all equal.

This limitation is not necessarily a negative quality. Ordinal scales may be sufficient if it is known that the classification variable possesses some arbitrariness. Thus, it may be useful to rank all 50 participants in an experiment with respect to some variable, and then compare, on some other variables or tests, the top 5 ranked participants with the bottom 5 ranked participants.

It has also been assumed in the previous discussion that the people or objects ranked received their respective rankings by a single classification variable. Of course, it is also possible to rank people or objects through more than one classification variable or even nebulous or hazy criteria. For example, movie critics are frequently known for presenting their rankings for the ten best movies or the ten worst movies. What are their classification variables? In this case, there are probably many classification or criteria variables, and some of them may even be inexplicable or unconscious. At the least, this may mean that some types of ordinal scales may be suspect.

Interval Scales. Interval scales probably receive the most statistical attention in the sciences. Interval scales give information about people or objects with respect to their ranking on some classification variable, and interpretations can be made with regard to how far apart the people or objects are on the variable.

With an interval scale, it is assumed the difference of a particular size is the same along all points of the scale. For example, on an attitude survey, the difference between the scores of 40 and 41 is the same as the difference between the scores of 10 and 11. On an interval scale it would also be assumed that a difference of 10 points between two scores would represent the same subjective difference anywhere along the scale. For example, it is assumed the difference between scores of 40 and 50 on an attitude survey is the same degree of subjective magnitude as between 5 and 15. Obviously, this is a crucial and difficult assumption to meet on any interval scale. Nevertheless, most measurement scales in the social sciences are assumed to be interval scales. Examples of interval scales are scores on intelligence tests, scores on attitude surveys, and most personality and psychopathology tests. However, there is substantial debate about whether these tests should be considered interval scales because it is questionable whether the equal magnitude assumption is met by these scales. In practice, however, most statisticians agree that the purported interval scales at least have the property of ordinal scales, and they may at least approximate interval scales. In a more practical evaluation, it is well documented these purported interval scales yield a plethora of interpretable findings.

Ratio Scales. Ratio scales have the properties of interval scales and, in addition, they have some rational zero point. This means that a zero point on a ratio scale has some conceptual meaning. It must be noted that in most of the social sciences, particularly in psychology or education, ratio scales are rarely used. Income could be measured on a ratio scale, because it makes sense to talk of 'zero' income, while it makes no sense to talk about 'zero' intelligence. Some types of psychological data might be measured on ratio scales, including such simple variables as weight or distance. In both of these latter cases, the zero value is a logical and conceptual place to begin the ratio measurement scale.

Although ratio scales may be thought of as the most sophisticated of the types of scales, they are not necessary to conduct research. Most types of statistical analyses and tests are designed to be used with interval scales. Indeed, one of the primary purposes of the science of statistics is to organize and understand groups of numbers and to make inferences about the nature of the world. Its purpose is not to create the perfect measuring device.

ROUNDING NUMBERS AND ROUNDING ERROR

The general rule for rounding numbers is if the number to be dropped is 5 or greater, then the remaining number is rounded up. If the number to be dropped is less than 5, then leave the remaining number unchanged. For example, when rounding to the nearest tenth, look at just the tenth and hundredth decimal places and ignore any places beyond (like thousandths):

10.977 would round to 11.0

125.63 would round to 125.6

100.059 would round to 100.1

6.555 would round to 6.6

6.5499 would round to 6.5

Anytime a number is rounded off, rounding error is introduced.

STATISTICAL SYMBOLS

Statisticians use symbols to represent various concepts. It will also be important for you to learn most of the common symbols they use. This should not be a very difficult task because the symbols are used over and over again, so you will have many opportunities to commit them to memory. Two of the most common symbols are $\bar{x}$ (pronounced x bar) which stands for the mean or average of a sample of numbers, and the Greek capital letter Σ (sigma) which is used to indicate the sum of a group of numbers. Let us practice with these two symbols. Find the average or $\bar{x}$ for the following group of numbers:

5,9,10

Intuitively, you probably know how to add the numbers up and divide by 3. Therefore, the mean (which is commonly used instead of average in statistics) is 8. However, let us use the word *set* instead of group. When you added each of the numbers together, you were taking the sum. Each number in the set of numbers can be represented by x. Therefore, in statistics, the formula for the mean would look like this:

$$\bar{x} = \frac{\Sigma x}{N}$$

where $\bar{x}$ = the mean of the average of the set of numbers

Σ = the sum of all the numbers in the set

N = how many numbers are in the set

$$\bar{x} = \frac{5 + 9 + 10}{3}$$

or

$$\bar{x} = \frac{24}{3}$$

$$\bar{x} = 8$$

Another common symbol is Σx^2 which says to square each number in the set and then add the results together. So, for the original set of numbers 5,9,10, obtain Σx^2.

$$\Sigma x^2 = 5^2 + 9^2 + 10^2$$

$$\Sigma x^2 = 25 + 81 + 100$$

$$\Sigma x^2 = 206$$

What does $(\Sigma x)^2$ indicate? Remember, in mathematics you must simplify what is in the parentheses first before you begin any other operations. So for this same set of numbers, add the numbers up first and then square this value.

$$(\Sigma x)^2 = (5 + 9 + 10)^2$$

$$(\Sigma x)^2 = (24)^2$$

$$(\Sigma x)^2 = 576$$

Does Σx^2 equal $(\Sigma x)^2$? By using the same numbers, you can verify that they are not equal! $\Sigma x^2 = 5^2 + 9^2 + 10^2$ or $\Sigma x^2 = 25 + 81 + 100$ which means $\Sigma x^2 = 206$. On the other hand $(\Sigma x)^2 = (5 + 9 + 10)^2$ or $(24)^2$ which means $(\Sigma x)^2 = 576$. Thus, you can see the resulting values are very different, and they cannot be used interchangeably. For those of you who may be weak in mathematics, it may be useful for you to note the following formulas carefully because you will need them later in this book. Summary: For the set of numbers 5,9,10 sum of the set

$$\Sigma x = 5 + 9 + 10$$

$$\Sigma x = 24$$

mean

$$\bar{x} = \frac{\Sigma x}{N}$$

$$\bar{x} = \frac{5 + 9 + 10}{3}$$

$$\bar{x} = 8$$

sum of each number in the set squared

$$\Sigma x^2 = 5^2 + 9^2 + 10^2$$

$$\Sigma x^2 = 25 + 81 + 100$$

$$\Sigma x^2 = 206$$

the squared sum of the set

$$(\Sigma x)^2 = (5 + 9 + 10)^2$$

$$(\Sigma x)^2 = (24)^2$$

$$(\Sigma x)^2 = 576$$

Sometimes the symbol x_i (sub i) appears. X_i refers to the ith number in the set, thus x_1 is the first number in the set, etc. Sometimes you will see x_n (x sub n), which stands for the nth number in the set or the last number in the set.

HISTORY TRIVIA

Although the history of mathematics is ancient, the science of statistics has a much more recent history. It is claimed that the first use of the word 'statistics' was by German Professor Gottfried Achenwall (1719–1792) in 1749, and he implied that statistics meant the use of mathematics in the service of the nation. For Achenwall, statistics was another word for state or political arithmetic which might include counting the population, the size of the army, or the amount and rate of taxation.

One of the earliest uses of the word 'statistics' in the English language was in 1791 by John Sinclair (1754–1835). In a preface to a survey of the population of Scotland, he acknowledged the political nature of statistics in Germany; however, he emphasized the use of statistics for social change and for the benefit of the people. His contributions would herald a slow but developing interest in the use of statistics to understand social problems.

Concurrent with these developments, mathematicians were making contributions in the area of probability theory that would ultimately form the foundations of inferential statistics. Pierre Simon, the Marquis de LaPlace (1749–1827) better known as 'LaPlace,' was a French mathematician who contributed much to early probability theory. Karl Friedrich Gauss (1755–1855) was a German mathematician and astronomer who also made valuable contributions to the foundations of both descriptive and inferential statistics.

In the middle to late 1800s, the application of statistics to social problems became more prominent. Adolph Quetelet (1796–1874), a Belgian mathematician and astronomer, extended some of Gauss' ideas to the analysis of crime in society. Francis Galton (1822–1911), cousin of Charles Darwin, was an English explorer, meteorologist, and scientist. His book, *Natural Inheritance*, published in 1889, has been recognized as the start of the first great wave of modern statistics. He also profoundly influenced another English person, Karl Pearson (1857–1936), who has been called the founder of the modern science of statistics. Pearson's intellectually provocative book, *The Grammar of Science*, was published in 1892. In it he stressed the importance of the scientific method to society and knowledge, and he believed statistical procedures were fundamental to the scientific method.

Florence Nightingale (1820–1910) was a contemporary of Karl Pearson and a friend of Francis Galton. She is typically remembered as a nurse and hospital reformer. However, she might just as well be remembered as the mother of descriptive statistics. She trained to become a nurse during the 1850s, and it was her strict observance to sanitation in hospitals that dropped death rates dramatically. She not only performed her nursing duties, but administrative duties as well. She also founded a school for the training of nurses and established nursing homes in England.

In the course of her administrative work, she developed a uniform procedure for hospitals to report statistical information about their patients. She is credited with developing the pie chart, which represents portions of the whole as pieces of a pie. She also argued to get statistics in the curriculum of higher education. She had suggested to Galton that a professorship be established for the statistical investigation of societal problems, and she pointed out issues to Galton which should be studied under the auspices of this professorship. These issues included crime, education, health, and social services. The University of London in 1911 finally established a Department of Applied Statistics, and appointed Karl Pearson its head with the title, Galton Professor.

Through the graphic representation of data, called a frequency histogram, Florence Nightingale convinced the Queen and the Prime Minister of England to establish a commission on the health and the care of the British Army. She did this by showing clearly with graphs that the rate of deaths for the military while at home in England was almost double the rate of equivalent non-military English males.

One of her biographers has argued Florence Nightingale's interest in statistics transcended her interest in health care, and was closely related to her strong religious convictions. She felt the laws governing social phenomena were also laws of moral progress; therefore, they were God's laws and could be revealed by the use of statistics.

KEY SYMBOLS AND TERMS

Descriptive Statistics – a group of techniques used to describe data in a straightforward manner, like tables, graphs, and pie charts.

Inferential Statistics – techniques that are used on samples in order to make inferences about population values.

Parameters – are common and conventionally accepted ways of measuring data characteristics.

Hypothesis – is an educated guess that guides research.

Sample – is a smaller group of scores selected from the population of scores.

Population – is most often a theoretical group of all possible scores with the same trait or traits.

Nominal Scale – a measurement scale in which the data are simply named.

Ordinal Scale – a measurement scale in which the data are rank ordered according to a trait.

Interval Scale – a measurement scale in which the units of measurement are equal along the length of the scale, but there is no rational zero point.

Ratio Scale – a measurement scale in which the units of measurement are equal, and there is a rational zero point.

Independent Variable – in an experiment, it is the variable that the experimenter manipulates. Also, it can be called the treatment variable or predictor variable.

Dependent Variable – in an experiment, it is a measure expected to vary across different levels of the independent variable. It is also called the response variable, or criterion variable in regression analysis.

Placebo Effect – it is the belief of the participant in an experiment that the independent variable will affect the participant's behavior. A placebo can also refer to an inert substance given to participants in the control group in order to control for placebo effects.

Double Blind Experiment – an experiment where the participants are kept unaware of the experimenter's hypothesis, and the experimenter is kept unaware of the participants' group affiliation (experimental or placebo group) until after the dependent variable has been measured.

Continuous Variable – a measurement scale where an individual measurement can be made at any point along the range of the scale.

Dichotomous Variable – a measurement scale where an individual measurement can only fall into two discrete categories.

Confounding Variable – is also called a nuisance variable. It is a variable that was not accounted for in the experimental design, varies systematically with the dependent variable, and prevents a clear interpretation of the independent variable upon the dependent variable.

1. Distinguish between a theory and a hypothesis.

2. Name two requirements to help make samples representative of the population.

3. Define dependent and independent variable. Give examples of each and explain how they relate.

4. Based on the following list, identify whether each item represents experimental control or statistical control.

   a. adding an additional group to an experiment

   b. increasing the number of participants in the study

   c. analyzing the data with a different statistical test

   d. using a double blind experiment

5. Which of the following are continuous variables?

   a. IQ   b.   eye color

   c. time in minutes   d.   gender

   e. height

6. Identify the type of measurement scale each of the following represent (*i.e.*, nominal, ordinal, interval, or ratio).

   a. weight

   b. distance

   c. birthplace

   d. heart rate

   e. IQ score

   f. 1st, 2nd, 3rd place finishers in a race

   g. race

   h. eye color

   i. 10 best psychology programs in America

7. Round each number to a whole number

   a. 10.999 c.   10.55

   b. 10.09 d.   11.399

8. Round each number to the nearest tenth.

   a. 12.988   c. 5.555

   b. 110.74   d. 55.549

9. Round each number to the nearest hundredth.

   a. 12.999   c. 6.055

   b. 225.433   d. 90.107

10. Find $\bar{x}$ for the following sets of numbers.

    a. 1, 2, 4, 5

    b. 8, 10, 15

    c. 7, 14, 22, 35, 40

11. Obtain Σx^2 for the following sets of numbers.

    a. 6, 9, 12

    b. 61, 75, 84, 85

    c. 100, 105, 110, 120

12. Obtain $(\Sigma x)^2$ for the following sets of numbers.

    a. 7, 9, 11

    b. 30, 41, 52, 63

    c. 225, 245, 255, 275

13. A psychologist is interested in determining whether a new 'anti-anx-iety' drug relieves anxiety in first year nursing students studying statistics. Fifteen students were assigned to one of three conditions: five students received a placebo, five received 50 mg of the new drug, and five received 100 mg of the drug. The students were then placed in a quiet room and were given 15 statistics problems to do. They were administered by the psychologist's assistant. After com-pleting the problems the papers were scored and the number of errors made by each student was used to determine their levels of anxiety.

    Based on the information provided above:

    a. Identify the dependent and independent variables.

    b. Determine the type of measurement scale used (interval, ratio, ordinal, nominal).

    c. Explain why the sample used is or is not representative of the population.

d. Using only the information above, give an operational definition of anxiety.

e. Define confounding variable and cite some examples that could create problems in this experiment.

14. The Governor of the State of Georgia in the USA recently proposed that the state should provide the parents of every Georgia newborn a classical music CD or cassette in order to raise the intelligence of the child. The Governor cited studies that have shown that college students who listen to classical music have higher IQ scores. What are some confounding variables in the Governor's proposal? What are some confounding variables in the college student studies?

2 Descriptive Statistics: Distributions of Numbers

Probably the oldest presentation of numbers in the history of descriptive statistics was the use of graphs and tables. In the late 1800s and early 1900s, the field of descriptive statistics consisted of mostly tabled numbers representing people's lifespans and other actuarial data. The presentation of numbers in graphs and tables is still very popular because people can still get a good and quick conceptual picture of a large group of numbers. As mentioned previously, Florence Nightingale impressed the Queen and the Prime Minister of England with her graph of death rates of British men versus British soldiers. Her graph, part of which is presented in Figure 2.1, is called a **bar graph**, and it is typically used with data based on nominal or ordinal scales. Florence Nightingale's nominal categories consisted of the British men and the British soldiers. The difference in their death rates can be seen by the differences in lengths between the two lines or bars.

Figure 2.1   **Relative mortality rates**

THE PURPOSE OF GRAPHS AND TABLES: MAKING ARGUMENTS AND DECISIONS

'Making decisions based on evidence requires the appropriate display of that evidence. Good displays of data help to reveal knowledge relevant to understanding mechanism, process and dynamics, cause and effect. That is, displays of statistical data should directly serve the analytic task at hand.'

– Edward Tufte, 1997

Based on her bar graph, Florence Nightingale was able to argue effectively to the Queen of England that unsanitary conditions in the English army led to higher death rates and that a national health commission should be established to improve living conditions. Her bar graph looks relatively simple and straightforward; however, it has a couple of deceptively powerful features. First, notice that she didn't actually present any evidence of the actual unsanitary practices of the British army, yet she was able to convince the Queen that it was true. She made an effective argument for cause and effect by graphing an appropriate variable (death rates/1000) relevant to her case. She had a hypothesis (an educated guess) that unsanitary conditions led to sickness and death. It is possible that had she chosen another variable like sickness rates, her argument may not have been as effective. Thus, choosing a variable relevant to her argument was one of her excellent decisions in preparing her graph. The second positive aspect of her simple bar graph was that she showed that death rates in the army were higher than of men the same age but not in the army. While this comparison may appear simple and obvious, it is a powerful lesson in making a clear argument. Florence Nightingale thought of two other explanations for the excessive death rates in the English army besides sanitation. Can you figure out what they were?

The most obvious objection (alternative hypothesis) she faced was that army life is inherently dangerous: wars kill people. It would be no wonder if death rates were higher for the British army. However, her bar graph very effectively dispelled this alternative hypothesis by making a relevant comparison. She graphed the death rates of **English army men at home and not at war compared to typical Englishmen**. By making this relevant comparison, she was able to show that it was not war-like conditions accounting for the higher death rates. Notice that this comparison didn't directly prove her argument for cause-and-effect but it did dispel a major rival explanation.

A second alternative hypothesis might have been that the age of the soldiers may have been the cause of the high death rates in the army. Perhaps, English soldiers were simply older than the typical Englishman, and thus, died at greater rates while still in England and not at war. Notice that there is some evidence for this argument if we examine her three bars for the typical Englishman. The death rates do appear to rise as age rises. However, Florence Nightingale brilliantly countered the age hypothesis by showing that English soldiers at home had higher death rates than Englishmen while comparing three different age groups, 20–25, 25–30, and 30–35. Again, Nightingale made a more effective argument for her hypothesis (unsanitary conditions) by **making relevant comparisons and by controlling for alternative hypotheses**.

During this same era in England, scores of people frequently died of cholera epidemics. Cholera, a disease still prevalent and deadly in the world today, comes from drinking water or eating food that has been contaminated by sewage. During the middle 1800s, the cause of cholera was still unknown, although there were at least two educated guesses, air or water transmission. There were also some more fantastical theories, for example, cholera was caused by vapors escaping from the burial grounds of plague victims (even though they had been dead and buried for over 200 years). In 1854, Dr. John Snow began investigating a cholera epidemic in London when over 500 people died within just ten days in one neighborhood. Snow's initial hypothesis that cholera was caused by contaminated water proved to be essentially correct. In order to make an effective argument for his hypothesis, Snow gathered evidence and made a graphic display. He accomplished this by getting a list of 83 officially recorded deaths within a short period of time. He plotted where these victims had lived on a map and discovered that a very large percentage lived near one particular well. After interviewing most of the families of the victims, he found that they did indeed get their drinking water from the Broad Street well. However, a rival argument was still plausible since not all of the 83 victims lived near the Broad Street well. Snow was able to fortify his hypothesis (and dispel a rival one) by his interviews because he found that, of the victims who lived nearer other wells, those victims had preferred the water from the Broad Street well, or went to a school that got its water from the Broad Street well. Within about a week, Snow presented his hypothesis and his graphic map display to the water authorities. They removed the pump handle from the Broad Street well, and the cholera epidemic quickly ended (see Tufte, 1997 for the complete and fascinating story).

Poorly conceived graphs and tables can also weaken arguments. The night of January 27, 1986, the makers of the space shuttle Challenger had a hypothesis that cold weather might make the rocket engine seals ineffective. It was predicted that the launch time temperature the next day would be about 27 degrees F. The average temperature of 24 previous launches was 70 degrees F. The lowest temperature of the previous 24 launches was 53 degrees F. and that launch had five serious mishaps related to seal failures, which was far more than any other launching. The shuttle manufacturers prepared 13 graphs and tables in a few hours that evening to support their hypothesis and faxed them to NASA. However, their 13 graphs and tables did not present their argument clearly. In one chart listing all of the prior rocket seal mishaps, there was no information about temperature. In another chart the same rocket was given three different names, making it difficult to determine which rocket had problems (yet there was only one rocket). Not one graph nor table simply listed the number of mishaps as a function of temperature, yet this information was present in the data. The information was not effectively extracted and presented. NASA, based on the 13 graphs and tables and two follow-up telephone conversations later that evening, was unconvinced that lower temperatures might affect the function of the seals. The next day the Challenger was launched, the rockets seals failed because of the cold weather, and the space shuttle blew up killing all seven crew members (Tufte, 1997).

'. . . *there are right ways and wrong ways to show data; there are displays that reveal the truth and displays that do not. And if the matter is an important one, then getting the displays of evidence right or wrong can possibly have momentous consequences.'*

– Edward Tufte, 1997

A SUMMARY OF THE PURPOSE OF GRAPHS AND TABLES

The contemporary statistician Edward Tufte (1997) nicely summarizes the reasoning behind gathering statistical evidence and statistical graphs and tables.

1. Document the sources of statistical data and its characteristics. Remember how Dr. Snow went to official death records? Not only was this method of gathering data more organized and official but the data he obtained through this method also provided him with standardized and very essential information like names, ages at death, and addresses where the victims lived. In the Challenger disaster, the rocket makers declined to put their individual names on the 13 charts and tables so ultimate responsibility remained anonymous. It might have been useful for officials at NASA to have been able to talk directly to some of the engineers who had the hypothesis about seal failure in cold weather. Furthermore, at the bottom of each chart the rocket makers placed a legal disclaimer that insinuated a kind of distrust for the chart makers and any of the charts' viewers. The moral here is that if one is going to make an argument for a hypothesis, particularly an argument in favor of safety, one should state the argument as strongly and effectively as possible. Anonymity and legal disclaimers do not make for an effective argument.

2. Make appropriate comparisons. Remember how Florence Nightingale controlled for an alternative hypothesis by restricting her comparisons to Englishmen the same age as English soldiers? And she further strengthened her argument by using English soldiers not at war but at home. By making relevant comparisons, she eliminated doubt about rival explanations.

Recently an 'answer column' in the newspaper was asked if there was a gender effect in developing dementia. 'Certainly,' was the answer, women are affected at a rate of three to six times that men are. This evidence was gathered by going to nursing homes and counting the number of demented males and females. The 'answer person' continued to speculate that since females have more of the hormone estrogen perhaps it had a deleterious effect. One glaring problem with this reasoning is that women live longer than men. If we count the absolute numbers of men and women in nursing homes, we will always be able to count more women than men. The 'answer person' failed to make a relevant comparison, that is, people at the same age.

Have you seen food store displays of 2% milk? What does the 2% represent? Do not feel badly if you are stumped. The 2% is supposed to represent the amount of fat in the milk. However, does 2% milkfat indicate that regular or whole milk has 100% milkfat or, in other words, does regular milk have 50 times the milkfat of 2% milk (2% = 50 times greater rate)? The 2% milk advertisements are a good example of the problems in interpreting comparisons. Regular milk has 8 grams of total fat in a one cup serving. In 2% milk there are 5 grams of total fat, thus, 2% milk has 62.5% of the total fat of regular milk. Therefore, 2% milk advertisements are misleading, either intentionally or inadvertently since few people would ever have imagined that 2% milk had 62.5% the fat of regular milk. This example of misadvertising again shows us the value in making relevant comparisons.

3. Demonstrate the mechanisms of cause and effect and express the mechanisms quantitatively. Many times it will not be sufficient to argue simply that we have discovered a real cause. The most effective argument for a causative hypothesis can be when we are able to demonstrate how varying the cause has a clear effect. Dr. Snow's hypothesis of causation was clearly supported when the water authorities removed the pump handle on the Broad Street well and death rates immediately declined. The demonstration of cause and effect had been clearly demonstrated through the mechanism of removing the pump handle.

In another highly visible display of the mechanism of cause and effect, a researcher proposed a few years ago that a bacteria was responsible for most ulcers. Despite some clinical evidence, there was much skepticism. In order to demonstrate a clear cause and effect relationship and in a highly visible display of the scientific method, the researcher had himself injected with a bacterial extract from a patient with an ulcer. He quickly developed an ulcer and furthermore cured it with antibacterial drugs. In this example, the researcher demonstrated one mechanism of cause and effect by injecting himself with the suspected bacteria and getting an ulcer. But he also demonstrated another mechanism of cause and effect consistent with his original hypothesis when he was able to cure himself of ulcers by using antibacterial drugs.

More recently a researcher claimed that HIV is not the cause of AIDS. With the same bravado, the researcher said he would put his controversial hypothesis to a similar test: he would inject himself with HIV. On the fateful day and before the media, the researcher did not show up.

4. Recognize the inherent multivariate nature of analytic problems. Most problems in science have a **multivariate** nature (more than one cause). For example, while most ulcers may have a bacterial origin some do not. Nor do all people exposed to the bacteria develop ulcers. In a recent newspaper headline, it was proposed that marriage tended to curb drug use. However, it is particularly true in psychology that there are multivariate causes. We are bombarded daily with overly simplistic explanations for behavior, like crime is caused by drug and alcohol abuse. The implication is that removing the cause (drugs and alcohol) removes its effects (criminal behavior). However, it is extremely rare in psychology, or in the rest of the sciences, that problems have a **univariate** nature (single cause). Eliminating drugs and alcohol from society will not decrease criminal behavior. In fact, there are some indications it might even increase. Forcing drug addicts to get married will not curb drug use. Severely addicted drug users will typically make terrible spouses and parents. What scientists can do, given the multivariate nature of most problems, is to argue clearly and effectively for some causal relationships while also remembering that nature is complex. Also, in many situations, varying causes may also vary in the strength of their contribution to a particular problem. Thus, criminal behavior may have a smaller contribution from heredity (they are born that way) and larger contributions from poverty, lack of education, and racial biases. Notice also that literally a hundred or more factors may be related to criminal behavior and that even when specifying a hundred factors, we still may not be able to predict accurately who will commit a crime. We continually read in the newspaper that *someone we would never suspect* has committed some heinous crime.

5. Inspect and evaluate alternative hypotheses. We saw that Florence Nightingale evaluated at least two rival hypotheses to her contamination hypothesis: age and war-like conditions. By making relevant comparisons to English soldiers at home and at various age groupings, she was able to dismiss both of them as plausible alternatives. Many times in scientific articles, researchers cannot evaluate and test for rival hypotheses. However, Tufte's suggestion to at least inspect other ideas may be useful. Many published scientific papers will simply note rival hypotheses in the introduction or discussion sections of their papers. If researchers 'save' the evaluation of rival hypotheses for the discussion section, they might do so by noting 'while there remains hypothesis A and hypothesis B for the present findings . . .' In this way science may be advanced, although the researcher has not formally evaluated the alternative hypotheses. Other researchers may then be able to generate research designs that may properly test rival ideas.

> 'When consistent with the substance and in harmony with the content, information displays should be documentary, comparative, causal and explanatory, quantified, multivariate, exploratory... . . . it also helps to have an endless commitment to finding, telling, and showing the truth.'
>
> *– Edward Tufte, 1997*

GRAPHICAL CAUTIONS

A note of caution is in order. People can just as easily fool themselves and others by bar graphs. An example of this tomfoolery is shown in Figure 2.2.

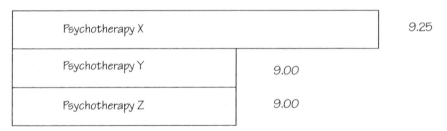

Figure 2.2

In this example, patient satisfaction for psychotherapy X appears to be far greater than for psychotherapy Y or psychotherapy Z (as measured on a scale of 1 to 10). In reality, the difference between the satisfaction rates for the three types of psychotherapy is very small (a quarter of one point on the scale). This means that the real mathematical differences in the bar lengths are not as great as the proponents of psychotherapy X have made them appear in the graph.

The following (Figure 2.3) graph tomfoolery occurred in a national truck advertisement.

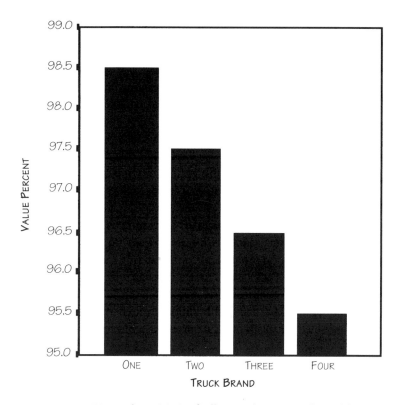

More than 98% of all Brand One trucks sold in the last ten years are still on the road.

Figure 2.3   **Percentage of trucks on the road after 10 years**

In this graph, the differences between the trucks, percentages have been magnified by cutting off the bottom 95% of the bars' heights. Although the graph makes it appear as if Truck one is much more reliable than any other truck (particularly that 'terribly' unreliable Truck four), the actual difference in reliability is barely 4%. And despite the graphical appearance of a major difference between Truck one and two, their reliability difference is less than 1%.

When continuous line scales are used to measure the dependent variable as in interval or ratio scales, as is common in behavioral sciences, a **frequency distribution** may be constructed. The frequency distribution is one of the most important graphic presentations in modern statistics. For example, let us imagine a shoe store owner who wishes to know what size shoes are available at any given point in time. The most inefficient way to present this data would be to write down all of the sizes in a single column. The problem with that approach is, the store owner would get a very limited idea of how many of each size of shoe existed. The frequency distribution, on the other hand, would give an immediate graphic or tabled picture of the shoes and their sizes. In addition, the frequency distribution can handle small or large samples. Suppose the inventory in the store consisted of one size 7, three size 5s, one size 3, two size 6s, and two size 4s. The first step in constructing a frequency distribution would be to arrange the shoe sizes from low to high in a table with their corresponding frequencies (how many of each) beside them.

Table 2.1

Shoe size	Frequency
three	1
four	2
five	3
six	2
seven	1

Note how easy it is now for the store owner to figure out how many of each shoe he or she has on hand. You can also imagine as the sample of shoes gets very large, this tabled frequency distribution will still be just as easy to understand.

Next let us construct a graphic picture of this frequency distribution. The graph will consist of two continuous line scales at right angles to each other which looks like this:

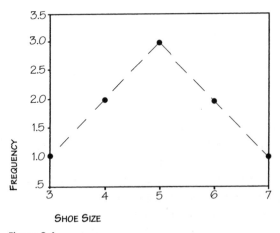

Figure 2.4

The horizontal axis most typically contains the line scale which measures the **dependent variable** or the thing that we are measuring. In this case the thing that we are measuring is shoe size. The vertical axis usually measures the frequency or how many of each shoe size exist within the sample. Note: each point in the graph represents an intersection of two lines drawn from each line scale. If you draw a line straight up from the lowest shoe size and draw a line straight across from the frequency of that shoe size, you will place a point at the intersection of these two lines. If you do this for each of the shoe sizes and you connect the points, it will generate a line which represents the frequency distribution. When the points are directly connected to one another with straight lines, this graph is also called a **frequency polygon**. A polygon is a closed plane figure having three or more straight sides. If we had represented the frequencies with bars as we did in a bar graph, the result would be called a **frequency histogram**. The difference between a bar graph and a histogram is simple: Bar graphs have spaces between the bars and histograms don't. Figure 2.5 presents the shoe data as a frequency histogram.

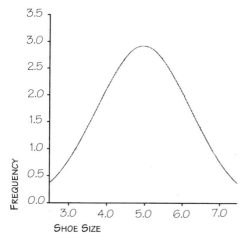

Figure 2.5

The general shape of this frequency polygon or histogram represents one of the most significant concepts in statistics. The shape resembles what is called the **normal curve** or **bell-shaped curve**. It can also be referred to as a normally distributed frequency distribution. In simple terms, it means when you are faced with a group of numbers representing most kinds of data, the resulting frequency distribution shows there are few cases which have a small amount of the dependent variable (the thing we are measuring, *e.g.*, low IQ, light weight, small shoe size). Most of the cases will have a medium amount of the dependent variable, and finally, just a few cases will have the largest amount of the dependent variable (the highest IQ, the heaviest weight, the largest shoe size). Not all kinds of data will result in a normally distributed frequency distribution. However, it is interesting that many kinds of data, including behavioral and biological, will produce a bell-shaped curve which approximates the normal distribution. The theoretical normal distribution is presented in Figure 2.6. The normal distribution also has special mathematical properties which will be discussed simply (and kindly) later.

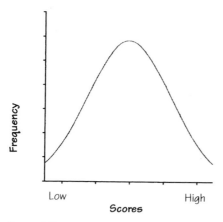

Figure 2.6

There are also common variations of the normal distribution. Sometimes, there are two more frequently occurring scores as measured by the dependent variable. This curve results in a bimodal distribution, as presented in Figure 2.7(A). If we use shoe size as an example, this would mean there is a large number of people who have a shoe size around 5 and an equally large number of people with a shoe size around 8. There are also two kinds of distributions which are variations on the normal distribution, and these are called **skewed distributions**. Figure 2.7(B) presents a **positively skewed distribution** (also called skewed right). Figure 2.7(C) presents a **negatively skewed distribution** (or skewed left).

Here in Colorado, my students remember whether a curve is positively skewed or negatively skewed by looking at the distributions as snowy mountains. In example B, if you are a 'normal' skier which side of the mountain would you ski down? You would ski to the right, so it is a distribution that is skewed to the right.

Many statistical software programs can calculate the skewness of a distribution. When you see a value of 0 for skewness, it means that the curve is not skewed, positive values indicate a right or positive skew, and negative values indicate a left or negative skew.

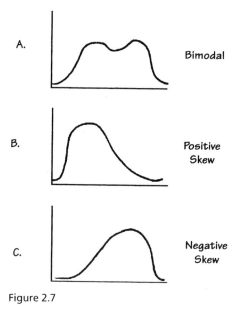

A. Bimodal

B. Positive Skew

C. Negative Skew

Figure 2.7

GROUPING DATA INTO INTERVALS

When we are dealing with a large group of numbers, or a group of numbers which are spread out over a large range of the dependent variable, we may wish to group the individual scores into categories or intervals. For example, let us look at the following set of scores on a personality test: 25, 27, 29, 30, 32, 36, 39, 44, 45, 47, 48, 48, 49, 52, 55, 56, 57, 63, 66, 67, 68. First, let us table the scores in a frequency distribution. Table 2.2 presents the resulting frequency distribution.

Table 2.2

Number	Frequency	Number	Frequency
25	1	48	2
27	1	49	1
29	1	52	1
30	1	55	1
32	1	56	1
36	1	57	1
39	1	63	1
44	1	66	1
45	1	67	1
47	1	68	1

Note that because the scores are spread out across the values of the dependent variable, a table of the data which simply lists frequency is relatively meaningless. This is also the case if a graph of the raw data is presented. Therefore, it may be better to group the scores together into intervals. Look at Table 2.3. The scores in the previous set of data have been grouped into intervals of 10.

Table 2.3

Interval	Frequency
20–29	3
30–39	4
40–49	6
50–59	4
60–69	4

By categorizing the data into intervals, we are now able to get a more meaningful picture. A graph of the grouped data appears in Figure 2.8.

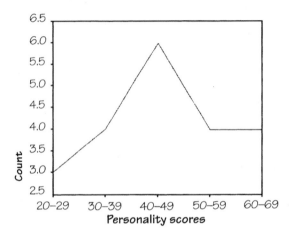

Figure 2.8

Now that we have grouped our data into intervals, the graphic presentation of the frequency distribution looks approximately mound-shaped.

ADVICE ON GROUPING DATA INTO INTERVALS

1. Choose interval widths that reduce your data to five to ten intervals. For example, if you have too few intervals, like two or three, then you may be crunching up your data too much. However, too many intervals may spread your data out too far. Generally, somewhere between five to ten total intervals seems to give a good picture of the data. A bad example appears in Table 2.2 where there are 20 intervals, and, thus, the data are too spread out. A better example appears in Table 2.3 where there are five intervals.

2. Choose the size of your interval widths based on understandable units, for example, in multiples of five or ten.

3. Make sure that your chosen intervals do not overlap. Look back at Figure 2.1. Notice that Florence Nightingale violated this rule. She grouped her males into ages 20 to 25, 25–30, and 30–35.

THE CUMULATIVE FREQUENCY DISTRIBUTION

In tabled frequency distributions, statisticians also use the concept of the cumulative frequency distribution. This parameter gives a picture of how many cases have been accounted for out of the total number of cases. Look at the example in Table 2.4.

Table 2.4

Shoe size	Frequency	Cumulative frequency
three	1	1
four	2	3
five	3	6
six	2	8
seven	1	9
Total	**9**	**9**

Note that at shoe size 3 and below, there is only one pair of shoes. At shoe size 4 and below, there are a total of three pairs of shoes (one pair of size 3 plus two pairs of size 4). This cumulative frequency continues until at size 7 and below, all nine pairs of shoes have been accounted for. Tabled frequency distributions are also often accompanied by the percentage of each individual score or the cumulative percentage. Look at Table 2.5.

Table 2.5

Shoe size	Frequency	Percentage	Cumulative percentage
three	1	11.1	11.1 (1/9)
four	2	22.2	33.3 (3/9)
five	3	33.3	66.7 (6/9)
six	2	22.2	88.9 (8/9)
seven	1	11.1	100.0 (9/9)
Total	**9**	**100.0**	

Note that at shoe size 3, there is a total of one pair of shoes out of the total of nine pairs. Therefore, one divided by nine is 11.1%. At shoe size 4, there are two pairs out of a total of nine pairs, and two divided by nine is 22.2%. In the cumulative frequency column, the cumulative percentages are totaled. Shoe size 3 accounts for 11.1% of all the shoes, and thus the total percentage of all shoe sizes at size 3 and below is also 11.1%. Shoe size 4 accounts for 22.2% of all the shoes sizes. The total cumulative percentage of shoe size 4 (and smaller) is 33.3% (obtained by 3/9). At shoe size 7 (and smaller), all nine pairs of the total nine pairs have been accounted for, therefore, the cumulative percentage is 100.0.

Let us return to the raw data presented in Table 2.2 and produce Table 2.6 which includes the percentage and cumulative percentage.

Table 2.6

Number	Frequency	Percentage	Cumulative percentage
25	1	5% (1/20)	5%
27	1	5%	10%
29	1	5%	15%
30	1	5%	20%
32	1	5%	25%
36	1	5%	30%
39	1	5%	35%
44	1	5%	40%
45	1	5%	45%
47	1	5%	50%
48	2	10%	60%
49	1	5%	65%
52	1	5%	70%
55	1	5%	75%
56	1	5%	80%
57	1	5%	85%
63	1	5%	90%
66	1	5%	95%
67	1	5%	100%
Total	20	100% (20/20)	

CUMULATIVE PERCENTAGES, PERCENTILES, AND QUARTILES

Cumulative percentages can be used to identify the position of a score in the distribution. Let us suppose that this previous set of scores was a scale measuring self-defeating behavior, where a high score indicated the strong presence of self-defeating attitudes, and a low score meant the absence of them. A raw score of 44 does not have much meaning, because its standing relative to the other scores is not known. However, the cumulative percentage shows that a raw score of 44 was in the lower half of all the scores, and 40% of all the subjects scored a 44 or below! Note: a raw score of 47 has 50% of all scores at that point or below.

Percentiles are derived from percentages, and they describe the score at or below which a given percent of the cases fall. The percentile scale is divided up into 100 units. Thus, a raw score of 47 is at the 50th percentile (The 50th percentile is also called the median of the distribution because it divides the distribution in halves). A raw score of 44 is at the 40th percentile.

Quartiles refer to specific points on the percentile scale. The first quartile refers to the 25th percentile, the second quartile refers to the 50th percentile, and the third quartile is the 75th percentile. Percentiles and quartiles are often used in educational measurement such as achievement testing.

STEM-AND-LEAF PLOT

In traditional frequency distributions, particularly when the data are plotted by intervals, each value of an individual score is lost. A contemporary statistician, Tukey (1977), created a **stem-and-leaf plot** which has a number of interesting features: it presents the data horizontally instead of vertically, it preserves each individual score, and extreme scores are readily observed.

In order to create a stem-and-leaf plot, let's use the data from Table 2.2. With each number, the leftmost digit will become the **stem** and the right digit becomes the **leaf**. The first number in this set is 25, so the left digit 2 will be the stem and the right digit 5 will be the leaf. The next two numbers in the set, 27 and 29 also share the same stem (2) but they have different leaves (7 and 9). Thus, a stem-and-leaf plot of the first three numbers in the data would look like this:

stem ⇒ 2 | 579 ⇐ leaves

Thus, attaching the stem (2) with each of its leaves (5, 7, and 9) gives us the original numbers 25, 27, and 29.

The complete stem-and-leaf plot of the data in Table 2.2 would look like this:

2 | 579

3 | 0269

4 | 457889

5 | 2567

6 | 367

Missing interval stems can also be presented. For example, what if the data in Table 2.2 did not have the numbers 63, 66, or 67 but had instead 70, 76, and 77. The stem-and-leaf plot would have looked like this:

2 | 579

3 | 0269

4 | 457889

5 | 2567

6 |

7 | 067

Notice how the interval stem 6 has no leaves. This indicates that there are no numbers in the set in the 60s.

For data with single digits, use a stem of 0. A stem-and-leaf plot of the shoe size data in Table 2.1 would look like this:

0 | 344555667

However, in this case the stem-and-leaf plot would not be very useful.

NON-NORMAL FREQUENCY DISTRIBUTIONS

When frequency distributions are graphically represented, sometimes the resulting line curve has varying symmetrical shapes, and sometimes it has asymmetrical shapes. Often, but not always, a frequency distribution will be mound-shaped. The shape of the mound is referred to as **kurtosis**. In Figure 2.9 there are four types of symmetrical distributions. Example A presents the normal frequency distribution or the bell-shaped curve. Example B has a pointed distribution. This tendency towards pointedness is referred to as **leptokurtosis**. Thus, example B presents a distribution which is leptokurtic. In example C, the distribution is flatter than the typical normal distribution. This tendency towards flatness is called **platykurtosis**, thus, the distribution in example C has platykurtic tendencies. A perfectly normal distribution is said to be mesokurtic.

Many statistical software programs can calculate the kurtosis of a distribution. When you see a value of 0 for kurtosis, it means that the curve is normal or mesokurtic, positive values indicate leptokurtosis, and negative values indicate platykurtosis.

There is one other famous symmetrical curve. It is called a **bimodal distribution**. It occurs in situations where there are two most frequently occurring scores but neither of the scores is at the exact center of the distribution. This distribution is presented in example D. Of course, it is also possible to imagine a trimodal distribution where there are three peaks in the distribution. One study in psychology found a trimodal distribution of the children's ages when they were first admitted to mental health care facilities (about ages 4 to 5, ages 7 to 8, and ages 10–12).

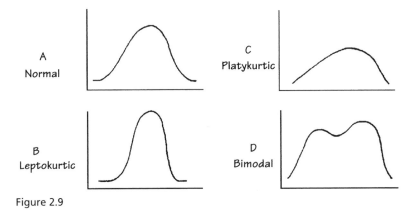

Figure 2.9

ON THE IMPORTANCE OF THE SHAPES OF DISTRIBUTIONS

The labels for the distributions may not have much meaning to you now. However, they are important, because they allow statisticians to communicate the shape of distributions quickly even without visual aids like graphs. In addition, the shapes of distributions of numbers are very important in inferential statistics where we will make inferences from samples of numbers about the populations from which they were drawn.

GOOD GRAPHS VERSUS BAD GRAPHS

The purpose of any graph should be to present data both accurately and conceptually clearly. With the proliferation of graphics programs for computers, undoubtedly there will be a proliferation of graphic presentations for data. However, it is already obvious that the most important purpose of statistics, that of conceptual clarity, is occasionally being forgotten in the midst of multicolor graphic options.

Therefore, it will be important for you to keep in mind some common pitfalls of graphic presentations. Watch for them as you prepare your own graphs, and beware of them while trying to interpret others' graphs.

Low Density Graphs. Tufte (1983) warns of low density graphs where very few data points are actually presented compared to the number of square inches taken up by the entire graph. A high density graph is not necessarily good, either. Remember, the purpose of a graph is to present the data accurately and clearly. Obviously, if there are very few data points to be presented, the readers may be better served by verbally presenting the data, instead of trying to impress readers with a graph.

Chartjunk. Tufte (1983) also warns of chartjunk which is an attempt to fill up the blank spaces of a graph with trivial or meaningless features. The tendency towards introducing chartjunk might be greater in low density graphs. Chartjunk may be simply unnecessary with a good graph. Indeed, it may be a sign of a bad graph. With the advent of computer graphic programs, it has become much easier to create graphs, and it has become much easier to fill graphs with meaningless or unnecessary features such as three dimensional bars or multidimensional bars and shading. While these additions are impressive, remember that the primary purposes of graphs are clarity of thought and efficiency of presentation.

Changing Scales Midstream (or Mid-Axis). Examine both axes of a graph. Make sure the scales used for each axis do not suddenly change. For example, if the horizontal axis was plotted by the years 1965, 1975, 1985, 1986, 1987, 1988, 1989, 1990, the grapher would have suddenly changed the measuring scale from 10 year periods to one year periods. Perhaps the purpose was to minimize differences in the graph, or perhaps it may have been to emphasize changes in the data. Either way, it was an unfair distortion of the data.

Labeling the Graph Badly. This can be done in a variety of ways. Frequently, when graphs are reproduced in print, they are reduced in size. Therefore, when you label parts of your graphs, make sure the labels are large enough to survive the reduction. Many advertisements violate this rule intentionally, such that restrictions to their offers appear in print so small that they are overlooked. Graphs should have clear, readable labels. The labels should not be ambiguous, incorrect, or illegible. All too often Professors create 'overhead' projections with the labels or explanations produced in standard typewriter font size. I have also witnessed slide presentations of graphs at conventions with the same labeling problem. Only those people in the very front row can read such small print and even to them, the small print is annoying.

The Multicolored Graph. The color option for computer graphic programs may introduce confusion into graphs instead of clarity. Although two or more colors may make bar graphs more artistically impressive, additional colors may just confuse the readers. Multicolors may also intentionally fool readers into thinking the graph is more meaningful than a simple black and white graph. Remember, conceptual clarity is a graph's most important purpose.

HISTORY TRIVIA

Who discovered the normal distribution? Some people claim Abraham De Moivre (1667–1754) has the clearest right to the discovery. He was born in France and raised in England. He was a mathematician, and it is known he gave private mathematics lessons in London. It is thought one of his students may have been the Englishman, Thomas Bayes (1702–1761), who went on to make important theoretical contributions to probability theory. De Moivre published two important works, one in 1711 and another in 1718. It is ironic that their largest appeal was not to mathematicians, but to gamblers, because the works dealt with games of chance. In fact, modern probability theory can trace its roots to letters of correspondence between famous mathematicians of the middle and late 1600s, discussing their attempts to solve and apply rules to gambling games. De Moivre is credited with developing the equation for the normal curve in approximately 1733.

Karl Friedrich Gauss (1777–1855), the German mathematician and astronomer, noticed that whenever large numbers of observations were made regarding the stars and planetoids, large numbers of errors always occurred. Gauss used the mathematical properties of the normal curve to develop a distribution of errors, and it became known as the Normal Law of Error. Quetelet (1796–1874), the Belgian astronomer and mathematician, may have been the first to develop an application of the normal curve other than describing a distribution of errors, instead using it to describe social and biological phenomena. However, it appears Francis Galton received most of the credit for turning the Gaussian Law of Error into a Law of Nature which is applicable to social and biological events.

Francis Galton (1822–1911), an English scientist and cousin of Charles Darwin, argued strongly that Gaussian errors were the exact opposite of what Galton felt should be studied. Gauss had argued these errors or deviations were to be removed or allowances were to be made for them. Galton claimed the errors or deviations were the very things he wanted to study or preserve! Galton published many books and articles, primarily on intelligence and inheritance. In 1876, he published a study of twins and the contributions of heredity and the environment and in it he coined the famous synonyms, 'nature' for 'heredity' and 'nurture' for 'environment.' Galton, within the next 10 years, developed the important statistical concept of correlation.

The application of the normal curve to social and biological phenomena is not without its critics. Jum Nunnally (1921–1986), an American professor of psychology, noted the distribution of psychological and educational test scores are seldom normally distributed, even if there are a large number of scores. He attributes this to the relationship each item on a test has to the others. Because it is expected the items have varying degrees of relationships to one another, the resulting distribution will be flatter (platykurtic) than the normal distribution. He notes that a perfectly normal distribution would be obtained only with 'dead data.' More recently Micceri (1989) surveyed 440 large-sample distributions of measures of achievement and psychological characteristics. He found that all 440 samples significantly deviated from the normal distribution. He likened the finding of a normal distribution of data to the probability of finding a unicorn.

It is interesting that Galton, even in his own time, recognized the potential limitations of the normal curve. In his biography he states he may have 'pushed the application of the Law of Frequency of Error somewhat too far.' However, consistent with modern thought, Galton believed, 'the applicability of that law is more than justified within . . . reasonable limits.'

Bar Graph – a graph which often represents nominal data in rectangular columns.

Frequency Distribution – a set of scores arranged in order of magnitude along the x-axis and the frequency of each score is represented along the y-axis.

Frequency Polygon – a graphic representation of a frequency distribution where the individual scores are grouped into class intervals.

Stem-and-Leaf Plot – a representation of a distribution where the individual scores are preserved.

Frequency Histogram – a graphic representation of a set of scores where the individual scores are represented by a bar whose height corresponds to the frequency of the score.

Normal Curve or Bell-Shaped Curve – the most frequently occurring distribution whose shape resembles a bell.

Skewed Distributions – an asymmetrical frequency distribution where either the right tail or the left tail is much longer than the other.

Positive Skew – an asymmetrical frequency distribution whose right tail is longer than the left.

Negative Skew – an asymmetrical frequency distribution whose left tail is longer than the right.

Kurtosis – refers to the peakedness or flatness of the overall shape of a frequency distribution.

Leptokurtosis – a frequency distribution that has a tendency towards peakedness.

Platykurtosis – a frequency distribution that has a tendency towards flatness.

Bimodal Distribution – a frequency distribution that has twin peaks.

Cumulative Frequency Distribution – a frequency distribution where the distribution of scores is progressively represented by the total frequency.

Percentiles – a distribution that is divided into hundredths.

Quartiles – a frequency distribution that is divided into fourths.

1. Name the four types of distributions.

2. The purpose of a graph is to present data _____
   and _____ .

3. The shape of the mound in a frequency distribution is referred to as
   _____ .

4. Leptokurtosis represents the tendency towards
   a. flatness   c. narrowness
   b. pointedness   d. both b and c

5. Using the following information construct a bar graph, and do not forget to label both axes of the graph. A psychologist in private practice is comparing her gross income for the first six months of the year. In January she made $3500; in February $5000; in March $2500; in April $3750; in May $4500; and in June $3900.

6. A psychologist is interested in IQs of ADHD children. The resulting scores were: 102, 115, 85, 100, 130, 86, 105, 114, 74, 83, 102, 101, 114, 93, 99,106,112, 97, 92, 94. Based on the psychologist's data:
   a. construct a frequency distribution. Use both a table and a graph.
   b. construct a stem-and-leaf plot.
   c. interpret the resulting data patterns

7. Describe the most likely shape for each of the following distributions (*i.e.*, normal, bimodal, positively skewed, or negatively skewed).
   a. The height for a large sample of men randomly selected from the general population.
   b. The height for a large sample of men and women randomly selected from the general population.
   c. The scores of clinical psychology Ph.D. candidates on the GRE psychology exam.
   d. Test scores for a group of average college students on an extremely difficult statistics final.

3 Statistical Parameters

MEASURES OF CENTRAL TENDENCY

In addition to graphs and tables of numbers, statisticians often use common parameters to describe sets of numbers. There are two major categories of these parameters. One group of parameters measures how a set of numbers is centered around a particular point on a line scale. This category of parameters is called **measures of central tendency**. The most famous and well-used statistical parameter from this category is the **mean** or **average**.

The Mean

The mean is the arithmetic average of a set of scores. There are actually different kinds of means, like the harmonic mean (which will be discussed later in the book) or the geometric mean. We will first deal with the arithmetic mean. The mean gives someone an idea about the center of a set of scores. The arithmetic mean is obtained by taking the sum of all the numbers in the set and dividing by the total number of scores in the set.

You probably already intuitively know how to obtain the mean. However, the following formula presents the mean in common statistical notation:

$$\bar{x} = \frac{\Sigma x}{N}$$

where

$\bar{x}$ = the mean

Σx = the sum of all the scores in the set

N = the number of scores or observations in the set

The mean has many important properties that make it useful. Probably its most attractive quality is that it has a clear conceptual meaning. People almost automatically understand and easily form a picture of an unseen set of numbers when the mean of that set is presented alone. Another attractive quality is that the mathematical formula is simple and easy. It involves only adding, counting, and dividing. The mean also has some more complicated mathematical properties which also makes it highly useful in more advanced statistical settings like inferential statistics. One of these properties is that the mean of a sample is said to be an **unbiased estimator** of the population mean. Remember that inferential statistics involves making inferences or guesses from a sample about a population. As an unbiased estimator, the mean of a sample has no tendency to overestimate or underestimate the population mean, μ (which is also written, *mu*, and pronounced *mew*). Thus, if consecutive random samples are drawn from a larger population of numbers, each sample mean is just as likely to be above μ as it is to be below μ. This property is also useful because it means that the formula for μ is the same as the formula for x̄. These formulae are as follows:

	Sample	Population
Mean	$\bar{x} = \dfrac{\Sigma x}{N}$	$\mu = \dfrac{\Sigma x}{N}$

The Median

Although the mean is the most widely used measure of central tendency, it is not always appropriate to use it. There may be many situations where the **median** may be a better measure of central tendency. The median value in a set of numbers is that value which divides the set into equal halves when all the numbers have been ordered from lowest to highest. Thus, when the median value has been derived, half of all the numbers in the set should be above that score and half should be below that score. The median is particularly appropriate when the distribution of numbers is skewed. In skewed distributions the mean is strongly influenced by the few high or low scores. For example, in a distribution which is positively skewed or skewed right, there are a few outlying high scores. The mean in this situation will be higher or pulled towards these high scores. Thus, the mean may not truthfully represent the central tendency of the set of scores because it has been raised by the few outlying high scores. In this case the median would be a better measure of central tendency. The formula for obtaining the median for a set of scores will vary depending on the nature of the ordered set of scores. The following two methods can be used in many situations:

Method 1: When the scores are ordered from lowest to highest and there is an odd number of scores, the middle value will be the median score. For example, examine the following set of scores:

> 7, 9, 12, 13, 17, 20, 22

Since there are 7 scores and 7 is an odd number, then the middle score will be the median value. Thus, 13 is the median score. In order to check whether this is true, look to see whether there are exactly the same number of scores above and below 13. In this case, there are 3 scores above 13 (17, 20, and 22), and there are 3 scores below 13 (7, 9, and 12).

Method 2: When the scores are ordered from lowest to highest and there are an even number of scores, the point midway between the two middle values will be the median score. For example, examine the following set of scores:

2, 3, 5, 6, 8, 10

There are 6 scores, and 6 is an even number; therefore, take the average of 5 and 6, which is 5.5, and that will be the median value. Notice that in this case, the median value is a hypothetical number that is not in the set of numbers. Let us change the previous set of numbers slightly and find the median:

2, 3, 5, 7, 8, 10

In this case 5 and 7 are the two middle values, and their average is 6; therefore, the median in this set of scores is 6. Let us obtain the mean for this last set, and that is 5.8. In this set the median is actually slightly higher than the mean. Overall, however, there is not much of a difference between these two measures of central tendency. The reason for this is that the numbers are relatively evenly distributed in the set. If the population from which this sample was drawn is normally distributed (and not skewed), then the mean and the median of the sample will be about the same value. In a perfectly normally distributed sample, the mean and the median will be the exact same value.

Now, let us change the last set of numbers once again:

2, 3, 5, 7, 8, 29

Now the mean for this set of numbers is 9 and the median remains 6. Notice that the mean value was skewed towards the single highest value (29), while the median value was not affected at all by the skewed value. The mean in this case is not a good measure of central tendency because 5 of the 6 numbers in the set fall below the mean of 9. Thus, the median may be a better measure of central tendency if the set of numbers has a skewed distribution. See Figure 3.1 for graphic examples.

If there are ties at the median value when you use either of the two previous methods, then you should consult an advanced statistics text for a third median formula which is much more complicated than the previous two methods. For example, examine the following set:

2, 3, 5, 5, 5, 10

There are an even number of scores and normally we would take the average of the two middle values. However, there are three 5s, and that constitutes a tie at the median value. Notice that if we used 5 as the value of the median, there is one score above the value 5 and there are two scores below 5. Therefore, 5 is not the correct median value.

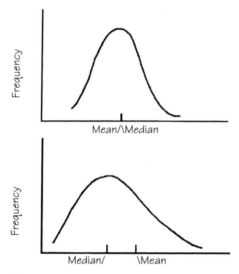

Figure 3.1

The Mode

The **mode** is a third measure of central tendency. The mode score is the most fre-quently occurring number in a set of scores. In the previous set of numbers, 5 would be the mode score because it occurs at a greater frequency than any other number in that set. Notice that the mode score in that set is 5 and the frequency (how many are there) of the mode score is 3 because there are three 5s in the set.

It is also possible to have two or more mode scores in a set of numbers. For example, examine this set:

2, 3, 3, 3, 4, 5, 6, 6, 6, 8

In this set there are two modes: one mode score is 3 and the other mode score is 6. The frequency of both mode scores is 3. A distribution which has two dif-ferent modes is said to be **bimodally** distributed. The mode score can change drastically across different samples, and thus, it is not a particularly good overall measure of central tendency. The mode probably has its greatest value as a meas-ure with nominal or categorical scales. For example, we might report in a study that there were 18 male and 14 female participants. Although it may appear obvi-ous and highly intuitive, we know that there were more male than female participants because we can see that 18 (males) is the mode score.

CHOOSING BETWEEN MEASURES OF CENTRAL TENDENCY

In one of my published journal articles, *Dreams of the Dying* (Coolidge & Fish, 1983), one of my undergraduates and I obtained dream reports from 14 dying cancer patients. In the method section of the article we reported the subjects' ages.

Rather than list the ages for all 14 subjects, what category of statistical parameters would be appropriate? Of course, it would be the category of measures of central tendency. The following numbers represent the subjects' ages at the time of their deaths:

28, 34, 40, 40, 42, 43, 45, 48, 59, 59, 63, 63, 81, 88

The mean for this set of scores is 52.4 and the median is 46.5 {(45 + 48)/2}. Typically a researcher would not report both the mean and the median, so which of the two measures would be reported? A graph of the frequency distribution (by intervals of 10 years) shows that the distribution appears to be skewed right. See Figure 3.2 for two versions of the frequency distribution.

Because of this obvious skew, the mean is being pulled by the extreme scores

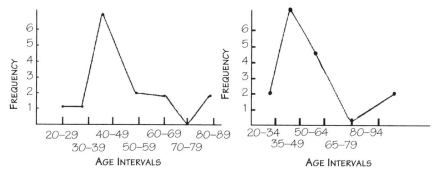

Figure 3.2

of 81 and 88. Thus, in this case we reported the median age of the subjects instead of the mean.

In most statistical situations, the mean is the most commonly used measure of central tendency. Besides its ability to be algebraically manipulated which allows it to be used in conjunction with other statistical formulae and procedures, the mean also is resistant to variations across different samples.

KLINKERS AND OUTLIERS

Sometimes when we are gathering data, we may have equipment failure, or in a verbal memory study we may have a subject who does not speak English. Datum from these situations may be simply wrong or clearly inappropriate. Abelson (1995) labels these numbers **klinkers**. Abelson argues that when it is clearly inappropriate to keep klinkers, they should be thrown out and additional data should be gathered in its place. Can I do this? Is this ethical? You may legitimately ask yourself these questions. Tukey (1969) and Abelson (1995) both warn against becoming too 'stuffy' and against the 'sanctification' of statistical rules. If equipment failure has clearly led to an aberrant score or a participant in an experiment failed to answer some questions because they did not speak English, then our way is clear. Keeping these data would ruin the true spirit of the statistical investigative process. We are not being statistically conservative but foolish if we keep such data.

The other type of aberrant score in our data is called an **outlier**. Outliers are deviant scores that have been legitimately gathered and are not due to equipment failures. While we may legitimately throw out klinkers, outliers are a much murkier issue. For example, the sports section of a newspaper reported that the Chicago Bulls were the highest paid team (1996–1997) in the National Basketball Association with an average salary of $4.48 million. However, let's examine the salaries and see if the mean is an accurate measure of the 13 Bulls players.

Table 3.1

Player	Salary
Jordan	$30,140,000
Rodman	$9,000,000
Kukoc	$3,960,000
Harper	$3,840,000
Longley	$2,790,000
Pippen	$2,250,000
Brown	$1,300,000
Simpkins	$1,040,000
Parish	$1,000,000
Wennington	$1,000,000
Kerr	$750,000
Caffey	$700,000
Buechler	$500,000

As a measure of central tendency, the mean salary is misleading. Only two players have salaries above the mean, Jordan and Rodman, while 11 players are below the mean. A better measure of central tendency for this data would be the median. Randy Brown's salary of $1,300,000 would be the median salary since six players have salaries above that number and six players are below that number. It is clear that Michael Jordan's salary is skewing the mean, and his datum would probably be considered an outlier. Without Jordan's salary, the Bulls average pay would be about $2,300,000, which is less than half of what the Bulls average salary would be with his salary. Furthermore, to show how Jordan's salary skews the mean, Michael Jordan makes about $3,000,000 more than the whole rest of the team's salaries combined!

We will revisit the issue of outliers later in the book. Notice in Jordan's case, we did not eliminate his datum as an outlier. First, we identified it as an outlier (because of its effect of skewing the mean of the distribution) and reported the median salary instead. Second, we analyzed the data (team salaries) both ways, with Jordan's salary and without, and we reported both statistical analyses. There is no pat answer to dealing with outliers. However, the latter approach (analyzing the data both ways and reporting both of the analyses) may be considered semi-conservative. The more conservative position would be to analyze the data and report it with the outlier and have no other alternative analyses without the outlier.

MEASURES OF VARIATION

The second major category of statistical parameters is **measures of variation**. Measures of variation tell us how far the numbers are scattered about the center-point of the set. They are also called measures of dispersion. There are three common parameters of variation, the **range**, **standard deviation**, and **variance**. While measures of central tendency are indispensable in statistics, measures of variation provide another important yet different picture of a distribution of numbers. For example, have you heard of the warning against trying to swim across a lake that averages only three feet deep? While the mean does give a picture that the lake on the whole is shallow, we intuitively know that there is danger because while the lake may average three feet in depth, there may be much deeper places as well as much shallower places in the lake. Thus, measures of central tendency are useful in understanding how scores cluster about a centerpoint, and measures of variation are useful in understanding how far, wide, deep, or high scores are scattered about the centerpoint.

The Range
The range is the simplest of the measures of variation. The range describes the difference between the lowest score and the highest score in a set of numbers. Typically statisticians do not actually report the range value but they do state the lowest and highest scores. For example, given the set of scores 85, 90, 92, 98, 100, 110, 122, the range value is $122 - 85 = 37$. Therefore, the range value for this set of scores is 37, the lowest score is 85, and the highest score is 122.

It is also important to note that the mean for this set is 99.6 and the median score is 98. However, neither of these measures of central tendency tells us how far the numbers are from this centerpoint. The set of numbers: 95, 96, 97, 98, 99, 103, 109 would also have a mean of 99.6 and a median of 98. However, notice how the range, as a measure of variation, tells us that in the first set of numbers they are widely distributed about the centerpoint (range = 37) while in the second set the range is only 14 ($109 - 95 = 14$). Thus, the second set varies less about their centerpoint than the first set.

Let us refer back to the ages of the dream subjects in the previously mentioned study. Although the mean of the subjects' ages was 52.4, we have no idea how the ages are distributed about the mean. In fact, because the median was reported, the reader may even suspect that the ages are not evenly distributed about the centerpoint. The reader might correctly guess the ages might be skewed; however, the reader would not know whether the ages were positively or negatively skewed. In this case the range might be useful. For example, it might be reported that the scores ranged from a low of 28 to a high of 88 (although the range value itself which is 60 might be of little conceptual use). Although the range is useful, the other two measures of variation, standard deviation and variance, are used far more frequently. The range is useful as a preliminary descriptive statistic; however, it is not useful in more complicated statistical procedures, and it varies too much as a function of sample size (the range goes up when the sample size goes up). The range also depends only on the highest and lowest scores (all of the other scores in the set do not matter), and a single aberrant score can affect the range dramatically.

The Standard Deviation

The standard deviation is a veritable bulwark in the sea of statistics. Along with the mean, the standard deviation is a theoretical cornerstone in inferential statistics. The standard deviation gives an approximate picture of the average amount each number in a set varies from the centerpoint. In order to appreciate the standard deviation, let us work with the idea of the average deviation. Let us work with a small subsample of the ages of the dream subjects:

28, 42, 48, 59, 63

Their mean is 48.0. Let us see how far each number is from the mean.

Each number (x_i)	Mean ($\bar{x}$)	Distance from mean ($x_i - \bar{x}$)
28	48	−20
42	48	−6
48	48	0
59	48	+11
63	48	+15

Note that the positive and negative signs tell us whether an individual number is above the mean or below the mean. The size or magnitude of the distance score tells us how far that number is from the mean.

In order to get the average deviation for this set of scores we would normally sum the five distance values and divide by 5 since there were five scores. In this case, however, if we sum −20, −6, 0, +11, and +15, we would get 0 (zero), and 0 divided by 5 is 0.

One solution to this dilemma would be to take the absolute value of each distance. This means that we would ignore the negative signs. If we now try to average the absolute values of the distances, we would obtain the sum of 20, 6, 0, 11, and 15, which is 52, and 52 divided by 5 is 10.4. Now, we have a picture of the average amount each number varies from the mean, and that number is 10.4.

The average of the absolute values of the deviations has been used as a measure of variation but statisticians prefer the **standard deviation** as a better measure of variation, particularly in inferential statistics. One reason for this preference, especially among mathematicians, is that the absolute value formula cannot be manipulated algebraically.

The formula for the standard deviation for a population is:

$$\sigma = \sqrt{\frac{\Sigma(x_i - \bar{x})^2}{N}}$$

Note that σ or sigma represents the population value of the standard deviation. You previously learned Σ as the command to sum numbers together. Σ is the capital Greek letter and σ is the lower case Greek letter. Also note that although they are pronounced the same, they have radically different meanings.

The sample standard deviation is a biased estimator of the population value, and consequently there is bad news and good news. The bad news is that there are two different formulae, one for the sample standard deviation and one for the population standard deviation. The good news is that statisticians do not often work with a population of numbers. They typically only work with samples and make inferences about the populations from which they were drawn. Therefore, we will only use the sample formula. The two formulae are presented as follows:

	Sample	Population
Standard Deviation	$S = \sqrt{\dfrac{\Sigma(x_i - \bar{x})^2}{N - 1}}$	$\sigma = \sqrt{\dfrac{\Sigma(x_i - \bar{x})^2}{N}}$

where S (capital English letter S) stands for the sample standard deviation.

CORRECTING FOR BIAS IN THE SAMPLE DEVIATION

Notice that the two formulae only differ in their denominators. The sample formula has $N - 1$ in the denominator and the population has only N. When it was determined that the original formula (containing only N in the denominator) consistently underestimated the population value when applied to samples, the correction -1 was added to correct for the bias. Note that the correction makes the numerator larger and that makes the value of the sample standard deviation larger (if we divide the numerator by a large number then it makes the numerator smaller, if we divide the numerator by a smaller number then that makes the numerator larger). The correction for bias has its greatest effect in smaller samples, for example, dividing by 9 instead of 10. In larger samples, the power of the correction is diminished yet statisticians still leave the correction in the formula even in the largest samples.

HOW THE SQUARE ROOT OF x^2 IS ALMOST EQUIVALENT TO TAKING THE ABSOLUTE VALUE OF X

As previously mentioned the absolute value method of obtaining the average deviation was used by early statisticians as a measure of variation. However, squaring a number and then taking the square root of that number also removes negative signs while maintaining the value of the distance from the mean for that number. For example, if we have a set of numbers with a mean of 4 and our lowest number in the set is 2, then $2 - 4 = -2$. If we square -2, we get 4, and the square root of $4 = 2$. Therefore, we have removed the negative sign and preserved the original value of the distance from the mean. Thus, when we observe the standard deviation formula, we see that the numerator is squared and we take the square root of the final value. However, if we take a set of numbers, the absolute value method for obtaining the standard deviation and the square root of the squares method will yield similar but not identical results. The square root of the square method has the mathematical property of weighting numbers that are farther from the mean more heavily. Thus, given the previous subset of numbers

28, 42, 48, 59, 63

the absolute value method for standard deviation (without the correction for bias) yielded a value of 10.4 while the value of the square root of the squares method is 12.5.

THE COMPUTATIONAL FORMULA FOR STANDARD DEVIATION

One other refinement of the standard deviation formula has also been made, and this change makes the standard deviation easier to compute. The following is the computational formula for the sample standard deviation:

$$S = \sqrt{\frac{\Sigma x^2 - \frac{(\Sigma x)^2}{N}}{N - 1}}$$

Remember that this computational formula is exactly equal to the theoretical formula presented earlier (the proof of their equality is presented in the appendix). The computational formula is simply easier to compute. The theoretical formula requires going through the entire data three times, once to obtain the mean, once again to subtract the mean from each number in the set, and a third time to square and add the numbers together. Note that on most calculators Σx and Σx^2 can be performed at the same time, thus the set of numbers will only have to be entered in once. Of course, many calculators can obtain the sample standard deviation or the population value with just a single button (after entering all of the data). You may wish to practice your algebra, nonetheless, with the computational formula and check your final answer with the automatic buttons on your calculator afterwards. Later in the course you will be required to pool standard deviations and the automatic standard deviation buttons of your calculator will not be of use. Your algebraic skills *will* be required so it would be good to practice them now.

The Variance

The variance is a third measure of variation. It has an intimate mathematical relationship with standard deviation. Variance is defined as the average of the square of the deviations of a set of scores from their mean. In other words we use the same formula as we did for the standard deviation except that we do not take the square root of the final value. The formulae are presented as follows:

	Sample	Population
Variance	$S^2 = \dfrac{\Sigma(x_i - \bar{x})^2}{N-1}$	$\sigma^2 = \dfrac{\Sigma(x_i - \bar{x})^2}{N}$

Statisticians frequently talk about the variance of a set of data, and it is an often used parameter in inferential statistics. However, it has some conceptual drawbacks. One of them is that the formula for variance leaves the units of measurement squared. For example, if we said that the standard deviation for shoe sizes is 2 inches, it would have a clear conceptual meaning. However, imagine if we said the variance for shoe sizes is 4 inches squared. What in the world does 'inches squared' mean? This conceptual drawback is one of the reasons that the concept of standard deviation is more popular in descriptive and inferential statistics.

THE USE OF THE STANDARD DEVIATION FOR PREDICTION

Pafrutti Tchebysheff (1821–1894), a Russian mathematician, developed a theorem which ultimately led to many practical applications of the standard deviation. Tchebysheff's theorem could be applied to samples or populations, and it stated that specific predictions could be made about how many numbers in a set would fall within a standard deviation or standard deviations from the mean. However, the theorem was found to be conservative, and statisticians developed the notion of the **Empirical Rule**. The Empirical Rule holds only for a normal distribution or relatively mound-shaped distributions.

The Empirical Rule predicts the following:

1. **Approximately 68% of all numbers in a set will fall within ± one standard deviation of the mean.**

2. **Approximately 95% of all numbers in a set will fall within ± two standard deviations of the mean.**

3. **Approximately 99% of all numbers in a set will fall within ± three standard deviations of the mean.**

For example, let us return to the ages of the subjects in the dream study previously mentioned:

28, 34, 40, 40, 42, 43, 45, 48, 59, 59, 63, 63, 81, 88

The mean is 52.4. The sample standard deviation computational formula is as follows:

$$S = \sqrt{\frac{\Sigma x^2 - \frac{(\Sigma x)^2}{N}}{N-1}}$$

$$S = \sqrt{\frac{42,287 - \frac{(733)^2}{14}}{13}}$$

$$S = \sqrt{\frac{42,287 - \frac{537,289}{14}}{13}}$$

$$S = \sqrt{\frac{42,287 - (38,377.7857)}{13}}$$

$$S = \sqrt{\frac{3,909.2143}{13}}$$

$$S = \sqrt{300.7088} = 17.34$$

Thus, S = 17.3

Now, let us see what predictions the Empirical Rule will make regarding this mean and standard deviation.

1. $\bar{x} + 1S = 52.4 + 17.3 = 69.7$

      $\bar{x} - 1S = 52.4 - 17.3 = 35.1$

Thus, the Empirical Rule predicts approximately 68% of all the numbers will fall within this range of 35.1 years old to 69.7 years old.

If we examine the data we find that 10 of the 14 numbers are within that range, and $^{10}/_{14}$ is about 70%. We find, therefore, that the Empirical Rule was relatively accurate but conservative.

2. $\bar{x} + 2S = 52.4 + 2 (17.3) = 52.4 + 34.6 = 87.0$

      $\bar{x} - 2S = 52.4 - 2 (17.3) = 52.4 - 34.6 = 17.8$

Inspection of the data reveals that 13 of the 14 numbers in the set fall within two standard deviations of the mean, or approximately 93%. The Empirical Rule predicted about 95%, thus, it was relatively accurate for this data, but this time it was too liberal a prediction.

3. $\bar{x} + 3\,S = 52.4 + 3\,(17.3) = 52.4 + 51.9 = 104.3$

$\bar{x} - 3\,S = 52.4 - 3\,(17.3) = 52.4 - 51.9 = 0.5$

All 14 of the 14 total numbers fall within three standard deviations of the mean. The Empirical Rule predicted 99%, and again we see the rule was relatively accurate and conservative.

Practical Uses of the Empirical Rule: IQ Tests

The Empirical Rule has great practical significance in the social sciences and other areas. For example, IQ scores (on Wechsler's IQ tests) have a theoretical mean of 100 and a standard deviation of 15. Therefore, we can predict with a reasonable degree of accuracy that 68% of a random sample of normal people taking the test should have IQs between 85 to 115.

Furthermore, only 5% of this sample should have an IQ below 70 or above 130 because the Empirical Rule predicted 95% would fall within two standard deviations of the mean. Because IQ scores are assumed to be normally distributed and both tails of the distribution are symmetrical, we can predict that 2.5% of people will have an IQ less than 70 and 2.5% will have IQs greater than 130.

What percentage of people will have IQs greater than 145? An IQ of 145 is exactly three standard deviations above the mean. The Empirical Rule predicts that 99% should fall within ± three standard deviations of the mean. Therefore, of the 1.0% who fall above 145 or below 55, 0.5% will have IQs above 145.

SOME FURTHER COMMENTS

The two categories of parameters, measures of central tendency and measures of variation, are important in both simple descriptive statistics and the higher level inferential statistics. As presented, you have seen how parameters from both categories are necessary to describe data. Remember that the purpose of statistics is to summarize numbers clearly and concisely. The parameters, mean and standard deviation, frequently accomplish these two goals, and a parameter from each category is necessary to describe data. However, being able to understand the data clearly is the most important goal of statistics. Thus, not always will the mean and standard deviation be the appropriate parameters to describe data. Sometimes the median will make better sense of the data, and most measures of variability do not make sense for nominal or categorical data.

HISTORY TRIVIA

Ronald A. Fisher (1890–1962) received an undergraduate degree in astronomy in England. After graduation he worked as a statistician and taught mathematics. At the age of 29 he was hired at an agricultural experimental station. Part of the lure of the position was that they had gathered approximately 70 years of data on wheat crop yields and weather conditions. The director of the station wanted to see if Fisher could statistically analyze the data and make some conclusions. Fisher kept the position for 14 years. Consequently, modern statistics came to develop some strong theoretical 'roots' in the science of agriculture.

Fisher wrote two classic books on statistics published in 1925 and 1935. He also gave modern statistics two of its three most frequently used statistical tests, t tests and analysis of variance. Later in his career, in 1954, he published an interesting story of a scientific discovery about eels and the standard deviation. The story is as follows:

Johannes Schmidt was an ichthyologist (one who studies fish) and biometrician (one who applies mathematical and statistical theory to biology). One of his topics of interest was the number of vertebrae in various species of fish. By establishing means and standard deviations for the number of vertebrae, he was able to differentiate between samples of the same species depending upon where they were spawned. In some cases, he could even differentiate between two samples from different parts of a fjord or bay.

However, with eels he found approximately the same mean and same large standard deviation from samples from all over Europe, Iceland, and Egypt. Therefore, he inferred that eels from all these different places had the same breeding-ground in the ocean. A research expedition in the Western Atlantic Ocean subsequently confirmed his speculation. In fact, Fisher notes, the expedition found a different species of eel larvae for eels of the eastern rivers of North America and the Gulf of Mexico.

KEY SYMBOLS AND TERMS

Measures of Central Tendency – parameters that measure the center of a frequency distribution.

Mean – is the arithmetical average of a group of scores.

Unbiased Estimator – a sample parameter that neither overestimates nor underestimates the population value.

Median – is the center of a distribution of scores such that half of the scores are above that number, and half of the scores in the distribution are below that number.

Mode – is the most frequently occurring score.

Measures of Variation – parameters that measure the tendency of scores to be close or far away from the center point.

Range – is the difference between the lowest score and the highest score in a distribution.

Standard Deviation – is a parameter of variability of data about the mean score.

Variance – a parameter of variability of data about the mean score and is the square of the standard deviation.

Empirical Rule – a rule of thumb that dictates that approximately 68% of all scores will fall within plus or minus one standard deviation of the mean in a normal distribution; 95% will fall within plus or minus two standard deviations; and 99% will fall within plus or minus three standard deviations of the mean.

1. Find the mean, median, and mode for each of the following sets of scores:

   Score   Mean   Median   Mode

   a.   35, 55, 80, 72, 55, 66, 74

   b.   110, 115, 102, 102, 107, 102, 108, 110

   c.   21, 19, 18, 30, 16, 30

   d.   24, 26, 27, 22, 23, 22, 26

2. In order to familiarize yourself with the measures of variation, compute the range, standard deviation, and variance for each set of scores.

a.	.6	.5	b.	258	335
	.8	.6		230	340
	.2	.3		305	250
	.9	.9			199
	.4	.7			

c.	9.62	9.30	d.	1001
	9.15	10.11		1253
	9.81	10.15		1234
	10.70	9.00		117
	9.45	10.63		

e.	.038	.030	f.	85.85
	.045	.046		79.20
	.071	.029		100.00
	.063			92.60
	.030			60.13

3. A group of college students was surveyed to determine the most popular type of music: Classical (C), Rock (RK), Soul (S), Rap (R), Heavy Metal (HM), or other (O). The final results were:

RK	R	O	R	RK
HM	R	R	O	R
RK	R	HM	S	HM
R	HM	C	HM	RK
RK	O	R	O	RK
R	RK	O	R	RK

a. Is it possible to find the mean, median, or mode?

b. Summarize the frequencies in a table.

4. The following is a list of scores on an anatomy exam. Find the mean, median, and mode for this class.

Student Grades

98	45	75	88
65	37	72	71
88	57	88	73
97	85	87	73
93	88	71	65
87	83	75	44
81	76	81	89

a. Mean   Median   Mode

b. What type of distribution does this represent?

c. Compute the standard deviation and variance.

d. Summarize the data in a stem and leaf plot.

5. Under what circumstances is it inappropriate to report the mean as the measure of central tendency?

6. The Full Scale IQ Score (FSIQ) for the Wechsler Adult Intelligence Scale – Revised (WAIS-R) has a mean of 100 and a standard deviation of 15. Based on this information use the empirical rule to answer the following. (Read each question carefully.)

a. What percentage of the population will fall within 1 standard deviation of the mean? Sketch a normal curve and shade in the corresponding area.

b.	What percentage of the population will fall within 2 standard deviations of the mean? Sketch a normal curve and shade in the corresponding area.

c.	What percentage of the population will have an FSIQ score above 130? Below 70?

4 Standard Scores, the Z Distribution, and Hypothesis Testing

STANDARD SCORES

A standard score is a number in a set where the mean and standard deviation of the set are already known or given to you. For example, an IQ score is a type of standard score because the mean for a population of subjects is 100 and the standard deviation is 15. A Scholastic Aptitude Test (SAT) score is also a type of standard score because its mean is known to be 500 and the value of one standard deviation is known to be 100. Many psychological tests, like the Minnesota Multiphasic Personality Inventory (MMPI), use T scores to interpret their clinical scales. A T score is one of the earliest types of standard scores whose mean is 50 and standard deviation is 10.

Standard scores are a useful way of summarizing and comparing scores across different kinds of tests. For example, if a patient's raw score on the Depression scale of the MMPI is 17, would that patient be considered depressed? Since we do not know the mean and standard deviation of the population of MMPI Depression scores, we cannot ascertain whether the patient is depressed or not. Furthermore, with only raw scores we cannot compare scores across different scales. For example, does this same patient's score of 17 on the Hypochondriasis scale of the MMPI mean that this patient has equal amounts of hypochondriasis and depression? The answer, of course, is no.

Imagine the advantages of reporting scores as standard scores. As soon as we know a patient's standard score, we would know immediately where they stand relative to a population measured on that same scale. A raw score of 17 on the MMPI Depression scale translates approximately to a T score of 50, thus, this patient is no more depressed than anyone in the normal population. However, a raw score of 17 on the Hypochondriasis scale would yield a T score of 65. This indicates that the patient is 1.5 standard deviations above the normative mean (since a standard deviation is 10 for T scores) or that the patient falls in about the 83rd percentile.

On many psychological tests, a T score of 70 (two standard deviations above the mean) is used to indicate a statistically defined level of psychopathology. Since the Empirical Rule states that approximately 95% of all scores will fall within plus or minus two standard deviations of the mean, then approximately

2.5% of the scores on the Depression scale of the MMPI would be above a T score of 70 (and 2.5% would fall below a T score of 30). On this statistical basis, psychologists decided that T scores above 70 on the clinical scales would be considered pathological or harmful. Would people who scored below a T score of 30 on the Depression scale be considered happy? No! Low scores on the MMPI are difficult to interpret. Frequently a low score means that the person is not taking the test honestly.

THE CLASSIC STANDARD SCORE: THE Z SCORE AND THE Z DISTRIBUTION

One important and classic type of standard score is the **z score**. The mean of a z distribution is 0.00 and the value of one standard deviation is 1.00. Z scores are the initial step in converting any raw score into any standard score. Most statisticians convert their raw scores into z scores and then into some other standard scores because z scores have two unfortunate characteristics. One is that a z score of 0 has bad connotations. For example, imagine if a person takes a major test at school and he/she comes home and reports getting a zero. People might automatically assume that the person did not do very well at all, yet if the person is reporting a z score, that person scored right at the mean. Z scores, thus, have a strong chance of being misinterpreted by people without statistical training. Indeed, even trained statisticians have unconscious difficulty with family members coming home and reporting zeros on tests.

A second problem with z scores is that they involve the use of negative numbers. Statisticians and non-statisticians have a natural aversion to negative numbers. We do not like them in our checking accounts, on final exams, or to identify football players ('Hey, look, there goes player number minus 76'). In the z distribution, any z score less than the mean of the z distribution will be a negative number. As shown previously when we derived the standard deviation formula, negative numbers are more difficult to deal with mathematically than positive numbers.

Despite the mathematical and psychological problems presented by negative z scores, the z distribution lies at the heart of inferential statistics. This is because statisticians use the z distribution to test experimental hypotheses, such as whether a drug is effective or not, or whether brain injured patients will score higher on some psychological test than a non brain injured group.

Let's briefly practice converting from T scores back into z scores. What would the earlier patient's Hypochondriasis T score of 65 be if it was reported as a z score? Since the value of one standard deviation of a T score is 10, and the patient scored 15 points above the T score mean of 50, then 15 divided by 10 is 1.5. Since the value of one standard deviation of a z score is 1.00, then we know the patient has a z score of 1.50 (or 1.50 standard deviations above the mean z score of 0). What about the patient's Depression score? Since their T score for the

Depression scale was 50 (which is the population mean for T scores), the equivalent z score would be 0.00 since that is the population mean of z scores. Furthermore, all of the following scores would be statistically equivalent since they would all indicate a value of one standard deviation above the mean: a z score of 1.00, a T score of 60, an IQ score of 115, and an SAT score of 600.

CALCULATING Z SCORES

In order to change scores into z scores, the mean and the standard deviation of the scores must be known (or at least derivable, meaning that we assume that we would have access to all of the scores in the set). If you have the mean and standard deviation for a set of scores, then any individual number can be converted to a z score by the following formula:

Z Score Formula

$$\text{z score} = \frac{x_i - \bar{x}}{S}$$

where

x_i = any individual score in the set of original numbers

$\bar{x}$ = the mean of the original set of numbers

S = the standard deviation for the original set of numbers

Even other standard scores can be converted into z scores or vice versa. In a previous example, a patient had a T score of 65 on the Hypochondriasis scale and a T score of 50 on the Depression scale. Now, let us use the formula for obtaining z scores to convert between the T scores and Z scores. For the math score:

$$\text{z score} = \frac{x_i - \bar{x}}{S} = \frac{65 - 50}{10} = \frac{15}{10} = 1.50$$

Thus, we see formulaically how the T score of 65 converts to a z score of 1.50. To convert the Depression score of 50 into a z score, we follow the same procedure:

$$\text{z score} = \frac{x_i - \bar{x}}{S} = \frac{50 - 50}{10} = \frac{0}{10} = -0.00$$

More Practice on Converting Raw Data into Z Scores

Let us convert the following raw scores, obtained by 10 students on a psychology exam, into z scores:

67, 74, 77, 81, 85, 89, 92, 93, 94, 99

First, obtain the mean and standard deviation. Note that these parameters should be accurate to at least 3 decimal places so that the resulting z score is accurate to at least 2 decimal places.

$$\bar{x} = \frac{\Sigma x}{N}$$

$$\bar{x} = \frac{851}{10}$$

$$\bar{x} = 85.1$$

$$S = \sqrt{\frac{\Sigma x^2 - \frac{(\Sigma x)^2}{N}}{N-1}}$$

$$S = \sqrt{\frac{73351 - \frac{724201}{10}}{9}}$$

$$S = \sqrt{\frac{73351 - 7420.1}{9}}$$

$$S = \sqrt{\frac{930.9}{9}}$$

$$S = \sqrt{103.433}$$

$$S = 10.17$$

Formula for z:

$$z = \frac{x_i - \bar{x}}{s}$$

Thus, for the raw score of 67:

$$z = \frac{67 - 85.1}{10.17} = -1.78$$

Raw score	Z score
67	−1.78
74	−1.09
77	−0.80
81	−0.40
85	−0.01
89	0.38
92	0.68
93	0.78
94	0.88
99	1.37

It is important to remember that the z is a standard score whose distribution is normally distributed. When the raw scores from any distribution are converted to z scores, the resulting distribution will approximate the normal distribution. This is because the z distribution is a perfectly normal distribution. By converting from raw scores to z scores, interpretations can then be made as if the scores are normally distributed. However, that is not to say that this assumption is correct. Perhaps, the raw scores come from a J curve. If raw scores were converted into z scores from a J curve distribution, the resulting inferences might be inappropriate. Fortunately, many raw score distributions in nature approximate the normal distribution. Therefore, the transformations from raw scores to z scores, in most cases, will be appropriate. How can you tell whether your data is normally distributed? Simply graph your data in a frequency distribution!

CONVERTING FROM Z SCORES TO OTHER TYPES OF STANDARD SCORES

Suppose that you wanted to convert the previous z scores into some other type of standard score. Once the z scores have been obtained, transforming them into any other type of standard score is easy, and can be performed with the following formula:

$$z' = \bar{x}' + (S')(z \text{ score})$$

where

z' (z prime) = the new standard score that you are trying to obtain (This is unknown in this equation. It is what you are trying to find!)

$\bar{x}'$ ($\bar{x}$ prime) = the mean of the new standard score (this is a known quantity)

S' (S prime) = the standard deviation of the new standard score (also a known quantity)

Thus, if the psychology professor wanted to convert the z scores into T scores ($\bar{x} = 50$, $S = 10$) in order to eliminate the bad connotations associated with the minus signs of the z scores, she would do the following:

Formula to convert from a z score to another type of standard score:

$$z' = \bar{x} + (S')(z \text{ score})$$

Thus, for the first score in the set, 67

$$T = 50 + (10)(-1.78)$$

$$T = 50 + (-17.8)$$

$$T = 50 - 17.8$$

T = 32.2

T = 32 (T scores are typically rounded to whole numbers)

Raw score	Z score	T score
67	−1.78	32
74	−1.09	39
77	−0.80	42
81	−0.40	46
85	−0.01	50
89	0.38	54
92	0.68	57
93	0.78	58
94	0.88	59
99	1.37	64

Note that the original raw scores give the students no indication of their standing relative to the class. If the professor graded without a curve, that is, a certain minimum score automatically earns a particular grade, and there is no limit on grades (the whole class can earn As), then the student should not care about a standing relative to the class. However, if the professor grades on a curve, where only the highest grades get As, the next highest get Bs, etc., then the conversion from raw scores to standard scores would be extremely meaningful for the student and professor.

THE Z DISTRIBUTION

Appendix A contains the z distribution. Table 4.1 has selected values of the z distribution, so that you can become comfortable reading the complete distribution.

Table 4.1   **Selected values of the Z distribution**

Positive z values	z	B	C
	0.00	.0000	.5000
	0.50	.1915	.3085
	1.00	.3413	.1587
	1.64	.4495	.0505
	1.96	.4750	.0250
	2.00	.4772	.0228
	2.57	.4949	.0051
	3.00	.4987	.0013
	4.00	.49997	.00003
Negative z values	−z	B	C

Appendix A shows the proportion of scores that can be expected from the mean of the z score (0.00) to a given z score. In fact the predictions that were made for the Empirical Rule were obtained from this z distribution. It can also be observed that the predictions of the Empirical Rule had rounding errors in them.

The shaded area B in Table 4.1 gives the proportion of scores from positive z scores of 0.00 (the mean of all z scores) to a target z score. The shaded area C in Table 4.1 gives the proportion of scores from the target z score to an infinitely high z score. Remember, a proportion can be read and interpreted just like decimals. In Table 4.1 look at a z score of 1.00, what is the B area proportion? It is .3413, and this means that .3413 of all z scores can be accounted for from a z score of 0.00 to a z score of 1.00. This proportion may also be converted to a percentage by moving the decimal place over to the right two places. Thus, 34.13% of all z scores occur from a z score of 0.00 to a z score of 1.00.

How many z scores can be accounted for from a z score of 0.00 to infinity? The total proportion of the z distribution is 1.00. A z score of 0.00 divides the z distribution exactly in half, thus, the answer to the previous question is .5000. Another way to solve this problem is to use the C area. Remember the C area gives the proportion of scores from the target z score to infinity. Thus, if our target z is 0.00 (because the question asked FROM a z score of 0.00 TO infinity) then the C area answers the question because the C area reports *from* the target z score (in this case 0.00) *to* infinity. The C area in this case is .5000.

Notice in Table 4.1 that the B area and the C area always add up to .5000. This makes sense because the B area gives the proportion from a z score of 0.00 to a target z, and the C area gives the proportion from the target z to infinity. Thus, one half of the z distribution can be accounted for by the B and C areas combined.

INTERPRETING NEGATIVE Z SCORES

The bottom part of Figure 4.1 shows what proportion of the z distribution is accounted for by negative z scores. A z score of −1.00 gives a B proportion of .3413 and a C proportion of .1587. Thus, from a z score of 0.00 to a z score of −1.00 accounts for .3413 or 34.13% of all the total scores. The C area shows what proportion of the scores are from the target z score (in this case −1.00) to an infinite negative z score, and that is .1587 or 15.87% of the total z scores.

TESTING THE PREDICTIONS OF THE EMPIRICAL RULE WITH THE Z DISTRIBUTION

The Empirical Rule stated that approximately 68% of all scores will fall within ± one standard deviation of the mean. If the proportion of scores from a z of 0.00 to a z score of 1.00 is .3413, then just double this proportional value to obtain the EXACT proportion of scores that fall within ± one standard deviation of the mean. The reason that this value is doubled is that we want to find the total proportion of scores from the mean z score of 0.00 to the value of one standard deviation on EACH side of the mean. This means from a z of 0.00 to a z score of 1.00, and from a z score of 0.00 to a z score of –1.00.

It is important to remember that the z scores can be positive or negative depending on whether the target scores are less than the mean or greater than the mean. Thus, the answer to the previous question would be .6826 or 68.26% of the total scores. The proportional answers must always be positive. Remember, the question in most cases will be what proportion of all cases can be accounted for by a z score. It would make no sense whatsoever to have a negative proportion or negative percentage of cases. The Empirical Rule gave only a rough or approximate estimate of the real proportion of scores. Remember the Empirical Rule stated ± two standard deviations from the mean would account for 95% of all the z scores? Let us check the accuracy of that statement.

A z score of 2.00 would be two standard deviations from the mean. The B area would have the proportion of scores from a z of 0.00 to a z of 2.00, and that is .4772. Doubling this value yields the value .9544 or 95.44%. Again, the proportion was doubled because we wanted to find the proportion of scores from a z score mean of 0.00 to two standard deviations on each side of the mean. Now we know that 95.44% of all z scores fall between a z score of –2.00 and +2.00, and the Empirical Rule stated 95%. Therefore, the Empirical Rule was incorrect by .0044 or .44%.

How accurate is the Empirical Rule at three standard deviations? It stated 99% of all scores would fall within three standard deviations of the mean. A z score of 3.00 yields a B proportion of .4987. If we add the proportions of z scores from a z score of 0.00 to a z of 3.00, and from a z of 0.00 to a z of –3.00, we obtain .9974 or 99.74% of all scores fall within ± three standard deviations of the mean on the z distribution.

WHERE ARE WE HEADING WITH ALL THIS INFORMATION?

Do not panic! It is too early for panic! Later in the course a few of you will be given permission to panic. The z distribution is a normal distribution. Approximately normal distributions occur more frequently than any other kind of distribution in nature. Throughout our exploration of inferential statistics we will assume (before we begin) that our data is normally distributed or, at the very least, that it is relatively mound-shaped. We will use the z distribution to test experimental hypotheses such as, 'Do copper bracelets reduce arthritic pain?'

HOW WE USE THE Z DISTRIBUTION TO TEST EXPERIMENTAL HYPOTHESES

Z scores are used as the foundation for the testing of experimental hypotheses. The following is a rough idea of how this process works. Let's imagine that we are trying to determine whether a new psychological treatment works better than nothing (a placebo). We will take all of the patients' scores in the experimental group and in the control group and convert them by a complicated formula into a single z score. Since 95.44% of all z scores should fall within plus or minus two standard deviations of the mean (–2.00 to +2.00) by chance alone, we will assume that only chance is operating if we obtain a z score within this range, that is, the new treatment does not work any better than a placebo. If all of the scores in the experiment result in a z score greater than +2.00 or less than a z score of –2.00, then it will be concluded that the experimental treatment works (since this would not likely be due to chance, or there would be about a 4.56% probability our results occurred by chance). See Figure 4.1 for a picture of these decisions based on a z distribution.

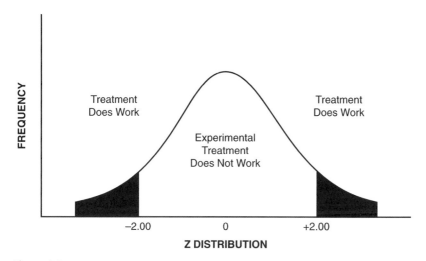

Figure 4.1

For example, with regard to the question, 'Do copper bracelets reduce the pain of arthritis?' it is automatically and statistically assumed that copper bracelets do NOT affect arthritic pain. An experiment is set up with an experimental group (wears a copper bracelet) and a control group (wears a placebo bracelet), and the two groups are measured for their respective numbers of pain complaints. The numbers are converted by a complicated formula to a z score (or a somewhat similar score). If the resulting z score is large (like greater than +2.00 or less than –2.00), then we can conclude that the treatment (copper bracelets) does affect the number of pain complaints. If the resulting z score is small (between –2.00 and +2.00), then it is concluded that the treatment does not affect the number of pain complaints any differently for the two groups.

MORE PRACTICE WITH THE Z DISTRIBUTION AND T SCORES

Example 1: Finding the area in a z distribution that falls above a known score where the known score is above the mean.

Question: What *percentage* of all T scores will be greater than a T score of 65?

Answer: First, you may find it highly useful to draw a picture of the question. See Figure 4.2.

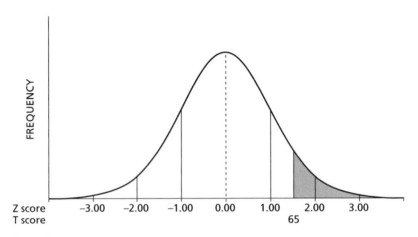

Figure 4.2

Next, convert the T score into a z score by the formula:

$$z = \frac{x_i - \bar{x}}{S} = \frac{65 - 50}{10} = \frac{15}{10} = 1.50$$

Now look in Appendix A and see what B and C areas are associated with a z score of 1.50. Do we want the B area or the C area? The B area represents the proportion of scores from a z of 0.00 to a z of 1.50. The C area represents the proportion of scores from a z of 1.50 to an infinite z. We want the C area that will give the proportion of scores greater than a z of 1.50, which is equivalent to a T score of 65. The proportion in the C area is .0668 or 6.68% of all T scores will be greater than a T score of 65.

Question: What percentage of all IQ scores will be greater than an IQ score of 117?

Answer: Draw a picture of the question. See Figure 4.3.

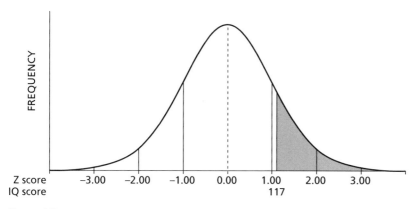

Figure 4.3

Next, convert the IQ score into a z score by the formula:

$$z = \frac{x_i - \bar{x}}{S} = \frac{117 - 100}{15} = \frac{17}{15} = 1.13$$

Now look in Appendix A and see what B and C areas are associated with a z score of 1.13. Do we want the B area or the C area? The B area represents the proportion of scores from a z of 0.00 to a z of 1.13. The C area represents the proportion of scores from a z of 1.13 to an infinite z. We want the C area that will give the proportion of scores greater than a z of 1.13, which is equivalent to an IQ score of 117. The proportion in the C area is .1292 or 12.92% of all IQ scores will be greater than an IQ score of 117.

Example 2: Finding the area in a z distribution that falls below a known score where the known score is above the mean.

Question: What percentage of all T scores will be less than a T score of 61?

Answer: It is highly useful to draw a picture of the question. See Figure 4.4.

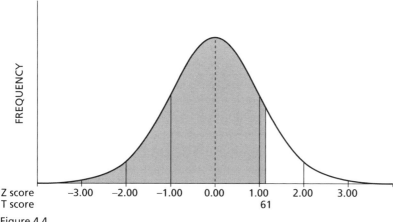

Figure 4.4

Next, convert the T score into a z score by the formula:

$$z = \frac{x_i - \bar{x}}{S} = \frac{61 - 50}{10} = \frac{11}{10} = 1.10$$

Now look in Appendix A and see what B and C areas are associated with a z score of 1.10. Do we want the B area or the C area? The B area represents the proportion of scores from a z of 0.00 to a z of 1.10. The C area represents the proportion of scores from a z of 1.10 to an infinite z. We want the B area that will give the proportion of scores greater than a z of 0.00 to a z of 1.10, which is .3643. Now all of the area under the z distribution less than a z of 0.00 is .5000. Thus, we add .5000 to .3643 which gives us a proportion of .8643 or 86.43% of all scores will be less than a T score of 61.

Question: What percentage of all IQ scores will be less than an IQ score of 117?

Answer: Again, you may find it useful to draw a picture of the question. See Figure 4.5.

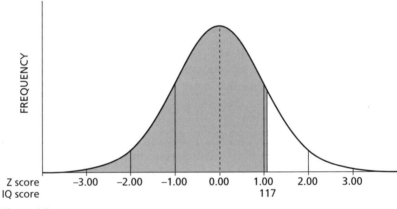

Figure 4.5

Next, convert the IQ score into a z score by the formula:

$$z = \frac{x_i - \bar{x}}{S} = \frac{117 - 100}{15} = \frac{17}{15} = 1.13$$

Now look in Appendix A and see what B and C areas are associated with a z score of 1.13. Do we want the B area or the C area? The B area represents the proportion of scores from a z of 0.00 to a z of 1.13. The C area represents the proportion of scores from a z of 1.13 to an infinite z. We want the B area that will give the proportion of scores from a z of 0.00 to a z of 1.13, which is 0.3708. The area of the z distribution less than a z of 0.00 is .5000. Again, we add .5000 to .3708 which gives us the proportion of 0.8708 or 87.08% of all scores will be less than a t score of 117.

Example 3: Finding the area in a z distribution that falls below a known score where the known score is below the mean.

Question: What *percentage* of T scores are less than a T score of 31?

Answer: First, draw a picture of the problem. See Figure 4.6.

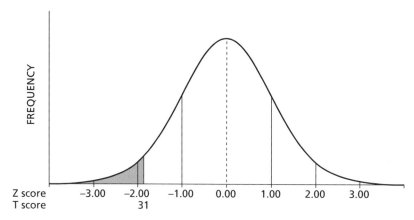

Figure 4.6

Next, convert the T score into a z score by the formula:

$$z = \frac{x_i - \bar{x}}{S} = \frac{31 - 50}{10} = \frac{-19}{10} = -1.90$$

Now look in Appendix A and see what B and C areas are associated with a z score of –1.90. Do we want the B area or the C area? The B area represents the proportion of scores from a z of 0.00 to a z of –1.90. The C area represents the proportion of scores from a z of –1.90 to an infinite negative z. We want the C area that will give the proportion of scores less than a z of –1.90. The proportion in the C area is .0287 or 2.87% of all T scores will be less than a T score of 31.

Question: What percentage of IQ scores are less than an IQ score of 70?

Answer: First, draw a picture of the problem. See Figure 4.7.
Next, convert the IQ score into a z score by the formula:

$$z = \frac{x_i - \bar{x}}{S} = \frac{70 - 100}{15} = \frac{-30}{15} = -2.00$$

Now look in Appendix A and see what B and C areas are associated with a z score of –2.00. Do we want the B area or the C area? The B area represents the proportion of scores from a z of 0.00 to a z of –2.00. The C area represents the proportion of scores from a z of –2.00 to an infinite negative z. We want the C area that will give the proportion of scores less than a z of –2.00. The proportion in the C area is .0228 or 2.28% of all IQ scores will be less than an IQ score of 70.

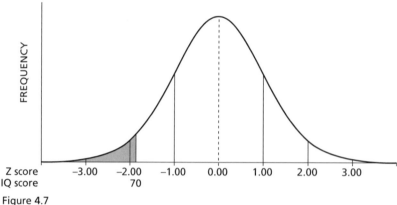

Z score   −3.00   −2.00   −1.00   0.00   1.00   2.00   3.00
IQ score   70

Figure 4.7

Example 4: Finding the area in a z distribution that falls above a known score where the known score is below the mean.

Question: What *percentage* of T scores will be greater than a T score of 31?

Answer: A picture of this problem is almost mandatory. It will help you to think clearly about the problem. See Figure 4.8.

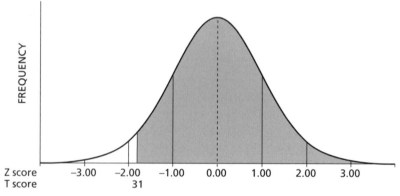

Z score   −3.00   −2.00   −1.00   0.00   1.00   2.00   3.00
T score   31

Figure 4.8

Next, convert the T score into a z score by the formula:

$$z = \frac{x_i - \bar{x}}{S} = \frac{31 - 50}{10} = \frac{-19}{10} = -1.90$$

Now look in Appendix A and see what B and C areas are associated with a z score of −1.90. Do we want the B area or the C area? The B area represents the proportion of scores from a z of 0.00 to a z of −1.90. The C area represents the proportion of scores from a z of −1.90 to an infinite negative z. This time we want the B area that will give the proportion of scores greater than a z of −1.90 to the mean z score of 0.00. This proportion in the B area is .4713. However, the question asked for the percentage of ALL T scores above 31 (not just to the mean z of

0.00). What proportion of all z scores are above a z of 0.00? The answer is .5000. Therefore, to obtain our final answer we have to add the B area proportion and .5000 together. Thus, we obtain the proportion of .9713 or 97.13% of all the scores are above a T score of 31.

Question: What percentage of IQ scores will be greater than an IQ score of 80?

Answer: A picture, please! See Figure 4.9

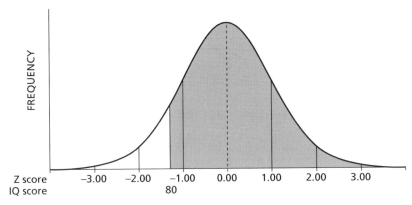

Figure 4.9

Next, convert the IQ score into a z score by the formula:

$$z = \frac{x_i - \bar{x}}{S} = \frac{80 - 100}{15} = \frac{-20}{15} = -1.33$$

Now look in Appendix A and see what B and C areas are associated with a z score of –1.33. Do we want the B area or the C area? The B area represents the proportion of scores from a z of 0.00 to a z of –1.33. The C area represents the proportion of scores from a z of –1.33 to an infinite negative z. This time we want the B area that will give the proportion of scores greater than a z of –1.33 to the mean z score of 0.00. This proportion in the B area is .4082. However, the question asked for the percentage of ALL IQ scores above 80 (not just to the mean z of 0.00). What proportion of all z scores are above a z of 0.00? Again, the answer is .5000. Therefore, to obtain our final answer we have to add the B area proportion and .5000 together. Thus, .4082 + .5000 = .9082 or 90.82% of all the scores are above an IQ score of 80.

Example 5: Finding the area in a z distribution that falls between two known scores where both known scores are above the mean.

Question: What percentage of T scores will be between a T score of 55 and a T score of 69?

Answer: First, draw picture of the problem. See Figure 4.10.

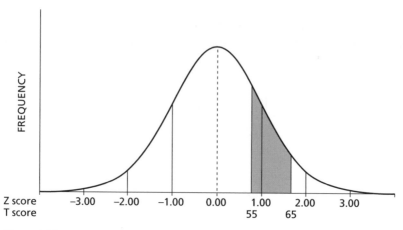

Figure 4.10

Our battle plan will be to convert both T scores to z scores, then we will look up the appropriate proportions, and subtract the smaller proportion from the larger proportion.

So after we draw a picture, convert the two T scores into z scores:

$$z = \frac{x_i - \bar{x}}{S} = \frac{55 - 50}{10} = \frac{5}{10} = 0.50$$

and

$$z = \frac{x_i - \bar{x}}{S} = \frac{69 - 50}{10} = \frac{19}{10} = 1.90$$

What B areas are associated with these two values of z? The B area associated with the z = 0.50 is .1915. This value represents the proportion of scores from a z of 0.00 to a z of 0.50. The B area for z = 1.90 is .4713.

Note from the picture that the B area associated with z = 1.90 includes the B area of the z = 0.50. Therefore, if we subtract out the smaller B area from the larger, the result will be the proportion of scores between these two values. Thus, .4713 – .1915 = .2798 or 27.98% of the total number of T scores fall between 55 and 69.

Question: What percentage of IQ scores will be between an IQ score of 105 and an IQ score of 125?

Answer: First, draw picture of the problem. See Figure 4.11.

Our battle plan will be to convert both IQ scores to z scores, then we will look up the appropriate proportions, and subtract the smaller proportion from the larger proportion.

So after we draw a picture, convert the two IQ scores into z scores:

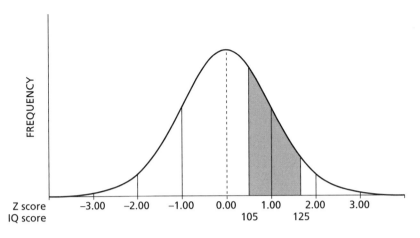

Figure 4.11

$$z = \frac{x_i - \bar{x}}{S} = \frac{105 - 100}{15} = \frac{5}{15} = 0.33$$

and

$$z = \frac{x_i - \bar{x}}{S} = \frac{125 - 100}{15} = \frac{25}{15} = 1.67$$

The B area for the $z = 0.33$ is .1293. This value represents the proportion of scores from a z of 0.00 to a z of 0.30. The B area for $z = 1.67$ is .4525. Subtracting the smaller proportion from the larger proportion will yield the proportion of IQ scores between 105 and 125. Thus, $.4525 - .1293 = .3232$ or 32.32% of the total IQ scores will be between an IQ of 105 and 125.

Example 6: Finding the area in a z distribution that falls between two known scores where one known score is above the mean and one is below the mean.

Question: What percentage of T scores will be between a T score of 43 and a T score of 72?

Answer: Again, draw a picture of the problem. See Figure 4.12.

Our battle plan here will be to convert both T scores to z scores, then we will look up the appropriate proportions and add the two proportions together.

Convert the two T scores into z scores:

$$z = \frac{x_i - \bar{x}}{S} = \frac{43 - 50}{10} = \frac{-7}{10} = -0.70$$

and

$$z = \frac{x_i - \bar{x}}{S} = \frac{72 - 50}{10} = \frac{5}{10} = 2.20$$

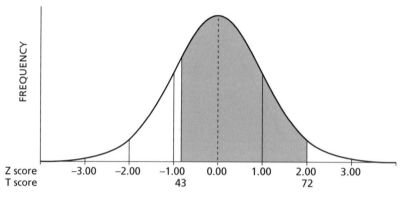

Z score   −3.00   −2.00   −1.00   0.00   1.00   2.00   3.00

T score   43   72

Figure 4.12

Do we want the B areas or the C areas? For the z score of –0.70, the B area is needed because we want the proportion of scores between the z = –0.70 and the mean z = 0.00. This proportion is .2580. For the z score of 2.20, the B area is also needed because we want the proportion from the mean z = 0.00 to z score = 2.20. This proportion is .4861. Next, simply add the two proportions together to get the total proportion, thus, .2580 + .4861 = .7441 or 74.41% of all the T scores fall between the T scores of 43 and 72.

Question: What percentage of IQ scores will be between a IQ score of 85 and a IQ score of 115?

Answer: Draw a picture. See Figure 4.13.

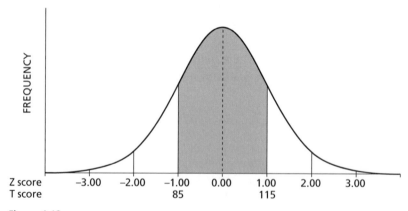

Z score   −3.00   −2.00   −1.00   0.00   1.00   2.00   3.00

T score   85   115

Figure 4.13

Here our battle plan will be to convert both IQ scores to z scores, then we will look up the appropriate proportions and add the two proportions together.
Convert the two IQ scores into z scores:

$$z = \frac{x_i - \bar{x}}{S} = \frac{85 - 100}{15} = \frac{-15}{15} = -1.00$$

and

$$z = \frac{x_i - \bar{x}}{S} = \frac{115 - 100}{15} = \frac{15}{15} = 1.00$$

The B area for the z = –1.00 is .3413. This value represents the proportion of scores from a z of 0.00 to a z of –1.00. The B area for z = 1.00 is also .3413. Adding these two proportions together will yield the proportion of IQ scores between 85 and 115. Thus, .3413 + .3413 = .6826 or 68.26% of the total IQ scores will be between an IQ of 85 and 115.

Example 7: Finding the area in a z distribution that falls between two known scores where both known scores are below the mean.

Question: What percentage of T score will be between a T score of 35 and 45?

Answer: Draw a picture of the problem. See Figure 4.14.

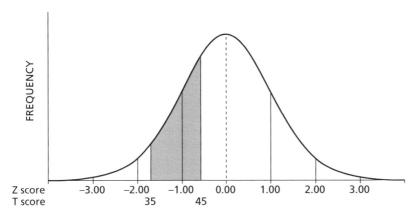

| Z score | –3.00 | –2.00 | –1.00 | 0.00 | 1.00 | 2.00 | 3.00 |
| T score | | 35 | 45 | | | | |

Figure 4.14

Our plan here will be to convert both T scores to z scores, then we will look up the appropriate proportions and subtract the smaller from the larger. Convert the two T scores into z scores:

$$z = \frac{x_i - \bar{x}}{S} = \frac{35 - 50}{10} = \frac{-15}{10} = -1.50$$

and

$$z = \frac{x_i - \bar{x}}{S} = \frac{45 - 50}{10} = \frac{22}{10} = -0.50$$

Do we want the B areas or the C areas? For the z score of –1.50, the B area is needed because we want the proportion of scores between the z = –1.50 and the mean z = 0.00. This proportion is .4332. For the z score of –0.50, the B area is also needed because we want the proportion from the mean z = 0.00 to z score

= .50. This proportion is .1915. Next, subtract the smaller proportion from the larger to get the remaining proportion, thus, .4332 – .1915 = .2417 or 24.17% of all the T scores fall between the T scores of 35 and 45.

Question: What percentage of IQ scores will be between an IQ score of 60 and an IQ score of 90?

Answer: First, draw a picture of the problem. See Figure 4.15.

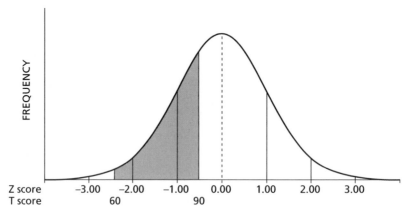

| Z score | -3.00 | -2.00 | -1.00 | 0.00 | 1.00 | 2.00 | 3.00 |
| T score | | 60 | | 90 | | | |

Figure 4.15

Our battle plan will be to convert both IQ scores to z scores, then we will look up the appropriate proportions and again subtract the smaller proportion from the larger proportion.

Convert the two IQ scores into z scores:

$$z = \frac{x_i - \bar{x}}{S} = \frac{60 - 100}{15} = \frac{-40}{15} = -2.67$$

and

$$z = \frac{x_i - \bar{x}}{S} = \frac{90 - 100}{15} = \frac{-10}{15} = -0.67$$

The B area for the z = –2.67 is .4962. This value represents the proportion of scores from a z of 0.00 to a z of –2.67. The B area for z = –0.67 is .2486. Subtracting the smaller proportion from the larger proportion will yield the proportion of IQ scores between 60 and 90. Thus, .4962 – .2486 = .2476 or 24.76% of the total IQ scores will be between an IQ of 60 and 90.

SUMMARIZING SCORES THROUGH PERCENTILES

People are fond of summarizing data with statistical parameters that are easily understood like **percentiles**. Percentiles provide a ranking in terms of a percentage compared to a larger group (presumably the population). One of the earliest uses of percentiles came from developmental psychology. In order to tell whether children were growing up normally, they were measured on height and weight charts. Parents were then told where their children stood compared to children the same age in terms of their height and weight. Let's practice determining percentiles using the previous examples in this chapter.

Question 1: A male child has a height of 4 ft. 4 inches at age 8. According to current height charts, the mean height at this age is 4 ft. 2 inches with a standard deviation of 3 inches. What is this child's height in percentile ranking?

Answer 1: First, convert the child's height into a z score by the formula:

$$z = \frac{x_i - \bar{x}}{S}$$

$$z = \frac{52 - 50}{3} = \frac{2}{3} = 0.67$$

The B area of the z distribution for this z score is .2486. Since this child's z score exceeds not only this area from the mean ($z = 0.00$) but also all scores below the mean (an area of .5000), the total proportion of the scores exceeded by this child's height is $.2486 + .5000 = .7486$. In order to convert this proportion to a percentage, move the decimal place over two places and round, thus, the child's height is at the 75th percentile. In other words, this child's height exceeds 75% of all children his age yet the child's height is exceeded by 25% of all children the same age.

Question 2: A female child weighs 37 pounds at age 4. According to current weight charts, the mean weight at this age is 40 lbs. with a standard deviation of 4 lbs. What is this child's weight as a percentile ranking?

Answer 2: First, convert the child's weight into a z score by the formula:

$$z = \frac{x_i - \bar{x}}{S}$$

$$z = \frac{37 - 40}{4}$$

$$z = \frac{-3}{4} = -0.75$$

The B area of the z distribution for this z score is .2743. In this case, we need the C area since the child's weight is less than the normative mean. The C area for this z score is .2266. Thus, the child's weight exceeds .2266, 22.66%, or places the child in the 23rd percentile of children the same age and gender. This child's weight exceeds 23% of children her age yet the child's weight is exceeded by 77% of children the same age.

HISTORY TRIVIA

Karl Pearson (1857–1936) has been called the founder of modern statistics. Although z scores and the z distribution were the creation of another English person, William S. Gosset (1876–1937), Pearson is credited (along with Francis Galton) with the inauguration of the first great wave of modern statistical thought.

Pearson received his college degree in mathematics and subsequently studied law and was admitted to the bar. He then traveled to Germany to study physics, metaphysics, religion, and socialism. In 1884, at the age of 27, he returned to London to take a university teaching position in applied mathematics and mechanics. In 1889, Francis Galton published his famous and influential book, *Natural Inheritance*. Pearson was 'immensely excited' (Pearson's own words) by Galton's book, particularly by Galton's theoretical work in determining laws of inheritance and the measurement of variability.

In 1892 Pearson published his most provocative book, *The Grammar of Science*, in which he promoted the value of statistical analysis. Pearson's subsequent contributions to the modern science of statistics is monumental. He developed the correlation coefficient which measures the strength of association between two variables (and which bears his name, the Pearson Product-Moment Correlation Coefficient), and he developed a goodness-of-fit test, known as the chi-square, which determines how well a mathematical prediction fits some observed data. Even modern statistical language contains many words that were originally coined by Pearson including: array, biserial, r, chi square, coefficient of variation, kurtosis, leptokurtic, platykurtic, multiple correlation, partial correlation, normal curve, standard deviation, and the symbol *sigma* for the standard deviation.

Pearson also enjoyed the process of intellectual argument. He took on the medical establishment when he published a paper which said that contemporary medical treatment of tuberculosis did not reduce death rates. Of course, he made this argument only after examining real data. His recommendations may now be considered bizarre, in light of modern treatment methods. He thought people with a 'pedigree' towards tuberculosis should be forbidden to marry or have children. His thoughts on racial intelligence are also now deemed racist and pernicious. However, the value he placed on the scientific method and statistical analysis are the very tools with which modern theorists might fight racism or examine whether a drug treatment for AIDS might be effective.

Alfred Adler (1870–1937) (who is not a famous statistician) was a Neo-Freudian who was interested in how our earliest memories affected our later psychological development. Helen Walker (born 1891), a statistician and historian, relates the following anecdotal story of Karl Pearson's earliest memory: 'Well,' he said 'I do not know how old I was, but I was sitting in a high chair and I was sucking my thumb. Someone told me to stop sucking it and said that unless I did so the thumb would wither away. I put my two thumbs together and looked at them a long time. "They look alike to me," I said to myself, "I can't see that the thumb I suck is any smaller than the other. I wonder if she could be lying to me."'

Helen Walker notes that in the anecdotal memory Pearson rejected constituted authority, appealed to empirical evidence, had faith in his own interpretation of the observed data, and finally there was 'imputation of moral obliquity to a person whose judgment differed from his own.' Walker claims that these characteristics were prominent throughout Pearson's career.

Karl Pearson should also be touted for an additional contribution to statistics and, that is, his son Egon S. Pearson (1895–1980) who was to make his own unique and important contributions to statistics. Finally, there is the anecdotal story of an American, with a recent Ph.D. degree, who went to study statistics in England and met and studied with Pearson. He asked Pearson how he had time to write, study and compute so much. Pearson's reply was: 'You Americans would not understand,' 'but I never answer a telephone or attend a committee meeting.'

KEY SYMBOLS AND TERMS

Standard Score – is a type of score whose mean and standard deviation are known or given.

Z Score – the first standard score whose mean is zero and whose standard deviation is 1.0.

T Score – a type of standard score with a mean of 50 and a standard deviation of 10.

Z Distribution – a distribution of scores according to frequency that is perfectly normally distributed based on z scores.

1. What is the definition of a standard score?

2. What are the advantages of using standard scores?

3. What is the mean and standard deviation of a z distribution? What are the disadvantages of using z distributions?

4. Using the Z Distribution table, find the proportion of the total area for the following, draw a picture of each, and shade in the target area.

   a. between a z score of 0 and a z score of 1.64

   b. between a z score of 0 and a z score of 1.96

   c. above a z score of 1.64

   d. above a z score of 1.96

   e. between a z score of 0 and a score of –1.00

   f. between a z score of 0 and a z score of –1.50

   g. below a z score of –1.64

   h. below a z score of –1.96

   i. below a z score of 0

   j. between a z score of –1.00 and a z score of 1.00

   k. between a z score of –1.96 and a z score of 1.96

   l. between a z score of –1.64 and a z score of 1.96

   m. between a z score of –2.00 and a z score of –1.00

   n. between a z score of –2.00 and a z score of 2.00

5. Using the Z Distribution table, find the z score(s) that most closely corresponds to the area described in each of the following, draw a picture of each, and shade in the target area.

   a. upper .2500   b. upper .1000

   c. upper .5000   d. lower .2500

   e. lower .0500   f. upper .0500

   g. upper .0100   h. upper .0010

   i. lower .5000   j. lower .7500

k. lower .9500   l. lower .9900

m. lower .0100   n. middle .5000

o. middle .9500   p. middle .9900

6. Convert the following patients' raw scores on the Coolidge Axis II Inventory (CATI) Depression scale into z scores and then into T scores:

    37, 42, 51, 38, 45, 50, 31, 44, 34, 39, 52, 40, 41, 36, 32

7. In the previous example, what proportion of the scores exceed a T score of:

    a. 70

    b. 60

    c. 50

8. In the previous example, determine what proportion of the scores are below a T score of:

    a. 65

    b. 60

    c. 40

9. In the previous example, determine what proportion of the scores are between:

    a. 40 and 60

    b. 30 and 70

    c. 45 and 55

5 Inferential Statistics: The Archetypal Experiment, Hypothesis Testing, and the Z Distribution

Inferential statistics is concerned with making conclusions about populations from smaller samples of the population. In descriptive statistics we were primarily concerned with simple descriptions of numbers by graphs, tables, and parameters that summarized sets of numbers like the mean and standard deviation. In inferential statistics our primary concern will be testing hypotheses on samples and hoping that these hypotheses, if true of the sample, will be true and generalize to the population. Remember that a population is defined as the mostly hypothetical group to whom we wish to generalize. The population is hypothetical for two reasons: First, we will rarely, if ever, have the time, money, nor will it be feasible to test everyone in the population. Second, we will attempt to generalize from a current sample to future members of the population. For example, if we were able to determine a complete cure for AIDS, we would hope that the cure would not only work for the current population of AIDS patients in the world but also any future AIDS patients.

The most common research designs in inferential statistics are actually very simple: We will test whether two different variables are related to each other (through correlation and the Chi-Square test) or whether two or more groups treated differently will have different means on a response (or outcome) variable (through t tests and analyses of variance). Examples of whether two variables are related to each other are plentiful in psychology and other disciplines. We might wish to know whether aging is related to depression, whether violent crime rates are related to crack cocaine use, whether breast implants are related to immuno-deficiency disease, whether twins' IQs are more highly related than siblings' IQs, etc. Note that finding a relationship between two variables does not mean that the two variables are causally related. However, sometimes determining whether relationships exist between two variables, like smoking and rates of lung cancer, may give up clues that allow us to set up experiments where causality may be determined. These experiments, typically with two or more groups treated differently, are the most powerful experimental designs in all of statistics. While correlational designs which determine whether two variables are related are very common and useful, they pale in comparison to the power of a well-designed experiment with two or more groups.

I have chosen to call the two group experiment the **archetypal experiment** for its universality, cross-culturality, and its power to determine causality. While some statisticians argue which statistical tests best evaluate the outcome of an experiment, few statisticians argue about the nature of the design of a classic two group experiment.

The archetypal experiment is a two group experiment, consisting of an experimental group and a control group. In this powerful experimental design, the independent variable is the factor which the experimenter is manipulating. For example, in a drug effectiveness experiment, the independent variable is the drug itself. One group receives the drug and the other group receives a placebo. The experimenter then determines whether the drug has an effect on some outcome measure, response variable, or as it is better known, the dependent variable. The dependent variable is the behavior which is measured or observed to change. The experimenter wants to see if the independent variable changes the dependent variable. Some statisticians have compared this experimental process to signal detection theory. If a treatment really works, then two groups treated differently should score differently on the chosen response variable, and this difference is the **signal**. If the treatment does not work at all, then the two groups should score similarly on the response variable. However, due to random errors or pure chance, it is highly unlikely that the two groups will have exactly the same means on the response variable. If the independent variable or treatment does not work, the two groups' means should be close but not exactly equal. This difference between the two means is attributed to chance or random error, and it is called noise. Thus, all of inferential statistics is based on the **signal-to-noise ratio**. If the independent variable really works, then the signal should be much greater than the noise. If the independent variable does not work, then the signal will not exceed the background noise.

HYPOTHESIS TESTING IN THE ARCHETYPAL EXPERIMENT

A hypothesis is an educated guess about some state of affairs. In the scientific world researchers are usually conservative about their results, and they assume that nothing has been demonstrated unless the results (signal) can be clearly distinguished from chance or random error (noise). Usually experiments are conducted with a research idea or hunch which is typically called the **research hypothesis**. However, in theory, all experiments are begun with a statement called the **null hypothesis** (abbreviated H_o) which states that there is no relationship between the independent variable and the dependent or response variable.

In the drug effectiveness experiment, the null hypothesis would be that the drug has no effect upon the dependent variable. Thus, frequently the null hypothesis will be the opposite of what the scientist believes or hopes to be true. The prior research hunch or belief about what is true is called the **alternative hypothesis** (abbreviated H_a).

As noted earlier in the book, science must work slowly and conservatively. The repercussions of badly performed science are deadly or even worse. Thus, the null hypothesis is usually a safe, conservative position which says that there is no relationship between the variables, or in the case of the drug experiment, that the drug does not affect the experimental group differently on the dependent variable compared to the control group.

HYPOTHESIS TESTING: THE BIG DECISION

All experiments begin with the statement of the null and alternative hypothesis. However, the null hypothesis is like a default position: we will retain the null hypothesis (or we will fail to reject the null hypothesis) unless our statistical test tells us to do otherwise. If there is no statistical difference between the two means, then the null hypothesis is retained. If the statistical test determines that there is a difference between the means (beyond just chance differences), then the null hypothesis will be rejected.

In summary, when a statistical test is employed, one of two possible decisions must be made: (a) retain the null hypothesis, which indicates that there are no differences between the two means other than chance differences, or (b) reject the null hypothesis, which indicates that the means are different from each other well beyond what would be expected by chance.

HOW THE BIG DECISION IS MADE: BACK TO THE Z DISTRIBUTION

A statistical test of the classic two group experiment will analyze the difference between the two means to determine whether the observed difference could have occurred by chance alone. The z distribution, or a similar distribution, will be used to make the decision to retain or reject the null hypothesis.

In order to appreciate how this occurs, imagine a large vat of 10,000 Ping-Pong balls. See Figure 5.1.

Let us suppose that each Ping-Pong ball has a z score written on it. Each z score on a ball occurs with the same frequency as in the z distribution. Remember that the z distribution reveals that exactly 68.26% of the 10,000 balls will fall within ± one standard deviation of the mean z score of 0.0. This means that 6,826 of the 10,000 Ping-Pong balls will have numbers ranging from –1.00 to +1.00. Also, 95.44% of all the balls will fall within ± two standard deviations of the mean. Therefore, 9,544 of the Ping-Pong balls will range between –2.00 and +2.00. Finally, we know that 9,974 Ping-Pong balls will be numbered from –3.00 to +3.00.

Now, let us play a game of chance. If blindfolded, I dig into the vat of balls and pull out one ball in random fashion, what is the probability that it will be a number between –1.00 and +1.00? If I bet you $20 that the number would be

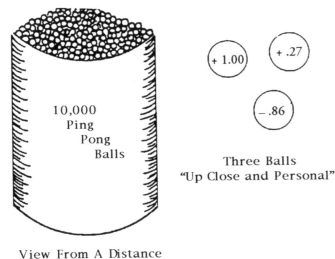

View From A Distance

Figure 5.1

greater than +1.00 or less than −1.00, would you take my bet? You should take my bet because the probability that the ball has a number between −1.00 and +1.00 is 68.26%. Therefore, you would roughly have a 68% chance of winning, and I would only have a 32% chance of winning.

How about if we up the stakes? I will bet you $100 that a z score on a randomly chosen ball is greater than +2.00 or less than −2.00. Would you take this bet? You should (and quickly) because now there is a 95.44% chance you would win and less than a 5% chance that I would win.

What would happen if we finally decided to play the game officially, and I bet a randomly chosen ball is greater than +3.00 or less than −3.00. You put your money next to my money. A fair and neutral party is chosen to select a ball and is blindfolded. What would be your conclusion if the resulting ball had a +3.80 on it?

There are two possibilities: either we both have witnessed an extremely unlikely event (only one ball out of 10,000 has a +3.80 on it), or something is happening beyond what would be expected by chance alone (namely that the game is rigged and I am cheating in some unseen way).

Now let us use this knowledge to understand the big decision (retain or reject the null hypothesis). The decision to retain or reject the null hypothesis will be tied to the z distribution. All of the individual subjects' scores in the two group experiment will be cast into a large and complicated formula and a single z-like number will result. In part, the size of this single z-like number will be based upon the difference between the two groups' means. If the two means are far apart, then the z-like number will most likely be large. If the two means are very close together (nothing but noise), then the z-like number will more likely be small. In other words, the data will be converted to a single number in a distribution similar to the z distribution. If this z-like value is a large positive or negative value (like +3.80 or −3.80), then it will be concluded that this is a low probability event. It is unlikely that what has happened was simply due to chance. In this case, the signal is much greater than the noise. Therefore, we will make the decision to reject the null hypothesis. If the formula yields a value between +1.96 and −1.96, then the null hypothesis will be retained because there is exactly a 95.00% probability by chance alone that the formula will yield a value in that range. See Figure 5.2 for a graphic representation of the z distribution and these decisions.

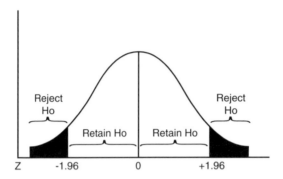

Figure 5.2

THE PARAMETER OF MAJOR INTEREST IN HYPOTHESIS TESTING: THE MEAN

In the classic two group experiment, the means of the two groups are compared on the dependent variable. The null hypothesis states that there is no difference between the two populations' means, such as:

$$H_o: \mu_1 = \mu_2$$

μ_1 represents the mean for the first population and μ_2 represents the mean for the second population. Since we will not be using the actual population means, we will be making inferences from our sample means, $\bar{x}_1$ and $\bar{x}_2$ to their respective population means. We hope that what we have concluded about the sample is true of the populations. Thus, we will be testing the sample means with the following null hypothesis:

$$H_o: \bar{x}_1 = \bar{x}_2$$

The alternative hypothesis is often what we hope is true in our experiment. The alternative hypothesis is most often stated as:

$$H_a: \mu_1 \neq \mu_2$$

However, in actuality the alternative hypothesis is stated in terms of the sample means,

$$H_a: \bar{x}_1 \neq \bar{x}_2$$

Note that the alternative hypothesis is stated as 'Mean 1 does not equal Mean 2.' This is its most common form, and it is called a **nondirectional alternative hypothesis**. Logically the 'does not equal' sign allows for two possibilities. One possibility is that Mean 1 is greater than Mean 2, and the other is Mean 1 can be less than Mean 2.

Since the archetypal experiment involves making inferences about populations, the analysis of the experiment involves inferential statistics. Thus, the mean is an essential parameter in both descriptive and inferential statistics.

NONDIRECTIONAL AND DIRECTIONAL ALTERNATIVE HYPOTHESES

An experimenter has a choice between two types of alternative hypotheses when hypothesis testing, a nondirectional or a directional alternative hypothesis. A **directional alternative hypothesis**, in the two group experiment, states the explicit results of the difference between the two means. For example, one alternative hypothesis could be:

$$H_a: \bar{x}_1 > \bar{x}_2$$

Here, the experimenter predicts that the mean for group 1 will be higher than the mean for group 2. Another possibility is that the experimenter predicts:

$$H_a: \bar{x}_1 < \bar{x}_2$$

Here, the experimenter predicts that Mean 1 will be less than Mean 2. In practice, however, most statisticians choose a nondirectional alternative hypothesis. One of the reasons for this is that the nondirectional alternative hypothesis is less influenced by chance. Directional alternative hypotheses, however, are not all bad. They are more sensitive to small but real differences between the two groups' means. Most statisticians agree that the directional alternative hypothesis should be reserved for situations where the experimenter is relatively certain of the outcome. It is legitimate to wonder, however, why the experimenter was conducting the experiment in the first place, if he or she was so certain of the outcome.

A DEBATE: RETAIN THE NULL HYPOTHESIS OR FAIL TO REJECT THE NULL HYPOTHESIS

Remember that the classic two group experiment begins with the statement of the null and the alternative hypotheses. Some statisticians are concerned about the wording of the decision that is to be made. Some say, 'The null hypothesis was retained.' Others insist that it should be worded, 'The null hypothesis was not rejected.' Although it may seem to be a trivial point, it has important implications for the entire meaning of the experiment.

After an experiment has been performed and statistically analyzed, and the null hypothesis was retained (or we failed to reject it), what is the overall conclusion? Does it really mean that your chosen independent variable has no effect whatsoever on your chosen dependent variable? Under any circumstances? With any kind of subjects? No! The conclusion is really limited to this particular sample of subjects. Perhaps, the null hypothesis was retained because your sample of subjects (although they were randomly chosen) acted differently than another or larger sample of subjects.

There are other possibilities for why the null hypothesis might have been retained besides the sample of subjects. Suppose that your chosen independent variable does affect your subjects but you chose the wrong dependent variable. One famous example of this type of error was in studies of the effectiveness of vitamin C against the common cold. Initial studies chose the dependent variable to be the number of new colds per time period (like per year). In this case the null hypothesis was retained. Does this mean that vitamin C has no effect on the common cold? No! When the dependent variable was the number of days sick within a given time period, the null hypothesis was rejected, and it was preliminarily concluded that vitamin C appears to reduce the number of days that people are sick with a cold.

It is important to remember that just because you do not find an effect does not mean it does not exist. You might be looking in the wrong place (using the wrong subjects, using the wrong experimental design) and/or you might be using the wrong dependent variable to measure the effect.

Thus, some statisticians recommend that it be stated, 'the null hypothesis was not rejected.' This variation of the statement has the connotation that there still may be a significant effect somewhere, but it just was not found this time. More importantly, it has the connotation that, although the null hypothesis was retained, it is not necessarily being endorsed as true.

THE NULL HYPOTHESIS AS A NON-CONSERVATIVE BEGINNING

The null hypothesis, however, is not always a conservative position. As in the case of sideeffects of most drugs, the null hypothesis is that there are no side effects of the drug! In order to correct for this unusual and non-conservative position, the experimenter might increase the regular dosage to exceptionally high levels. If no harmful side effects were observed, then it might be preliminarily concluded that the drug is safe.

Even in cases where high levels are shown to be safe, scientists are still conservative. Scientists will typically call for a **replication** of the experiment, which means that the experiment will be performed again by another experimenter in a different setting. In science it is said that one cannot PROVE anything. Even after high dosages are shown to have no side effects and after repeated experiments with other dosage levels, the drug's safety still has not been proven. Successful replication simply lends additional weight to the hypothesis that the drug is safe. It may still be found that the drug is not safe under other conditions or for other types of people. Thus, replication is important because it generally involves manipulations of other independent variables like the types of subjects, their ages, etc.

THE FOUR POSSIBLE OUTCOMES IN HYPOTHESIS TESTING

There are four possible outcomes in hypothesis testing: two correct decisions and two types of error.

1. Correct Decision: Retain H_o, when H_o is actually true.

In this case, we have made a correct decision. In an example involving the relationship between two variables, the H_o would be that there is no relationship between the two variables. A statistical test (like correlation) is performed on the data from a sample, and it is concluded that any relationship that is observed is due to chance. In this case we retain H_o and infer that there is no relationship between these two variables in the population from which the sample was drawn. In reality, we do not know whether H_o is true or not. However, if it is true for the population and we retain H_o for the sample, then we have made a correct decision.

2. Type One Error: Reject H_o, when H_o is actually true.

The **Type One Error** is considered to be the more dangerous of the two types of errors in hypothesis testing. When a person commits a Type One Error they are claiming that their research hypothesis is true, when it really is not true. This is considered to be a serious error because it misleads people. Imagine for example, a new drug for the cure of AIDS. A researcher who commits a Type One Error is claiming that the new drug works when it really does not work. People with AIDS are being given false hopes, and resources that should be spent on a drug that really works will be spent on this bogus drug. The probability of committing a Type One Error should be less than 5 chances out of 100 or $p < .05$. The probability of committing a Type One Error is also called **alpha**.

3. Correct Decision: Reject H_o, when H_o is actually false.

In this case, we have concluded that there is a real relationship between the two variables and it is probably not due to chance (or that there is a very small probability that our results may be attributed to chance). Therefore, we reject H_o and assume that there is a relationship between these two variables in the population. If in the population there is a real relationship between the two variables, then by rejecting H_o we have made the correct decision.

4. Type Two Error: Retain H_o, when H_o is actually false.

A **Type Two Error** occurs when a researcher claims that a drug does not work when in reality, it does work. This is not considered to be as serious an error as the Type One Error. Researchers may not ever discover anything new or become famous if they frequently commit Type Two Errors, but at least they have not misled the public and other researchers. The probability of a Type Two Error is also called **beta**.

A summary of these decisions appears in Table 5.1.

Table 5.1

Our decision	In reality	The result
Retain H_o	H_o is true	Correct Decision
Reject H_o	H_o is true	Type One Error (alpha = α)
Reject H_o	H_o is false	Correct Decision
Retain H_o	H_o is false	Type Two Error (beta = β)

SIGNIFICANCE LEVELS

A test of significance is used to determine whether we retain or reject H_o. The significance test will result in a final test statistic or some single number. If this number is small, then it is more likely that our results are due to chance, and we will retain H_o. If this number is large then we will reject H_o and conclude that there is a very small probability that our results are due to chance. The minimum conventional level of significance is p (or alpha) = .05. This final test statistic is compared to a distribution of numbers which are called **critical values**. The test statistic must exceed the critical value in order to reject H_o.

SIGNIFICANT AND NONSIGNIFICANT FINDINGS

When significant findings have been reported in an experiment, it means that the null hypothesis has been rejected. The word **'nonsignificant'** is the opposite of significant. When the word nonsignificant appears, it means that the null hypothesis has been retained. Do not use the word **'insignificant'** to report nonsignificant statistical findings. Insignificant is a value judgment, and it has no place in the statistical analysis section of a paper.

In the results section of a research paper, significant findings are reported if the data meet an alpha level of .05 or less. If the findings are significant, it is a statistical convention to report them significant at the lowest alpha level possible. Thus, although H_o is rejected at the .05 level (or less), researchers will check to see if their results are significant at the .01 or .001 alpha levels. It appears more impressive if a researcher can conclude that the probability that their findings are due to chance is $p < .01$ or $p < .001$. It is important to note, this does not mean that results with alphas at .01 or .001 are any more important or meaningful than results reported at the .05 level.

TRENDS AND DOES GOD REALLY LOVE THE .05 LEVEL OF SIGNIFICANCE MORE THAN THE .06 LEVEL?

Sometimes researchers will report 'trends' in their data. This usually means that they did not reject H_o but that they came close to doing so. For example, computers do many of the popular statistics, and they commonly print out the exact alpha levels associated with the test statistic. A trend in the data may mean that the test statistic did not exceed the critical value at the .05 level, but the findings may be associated with an alpha of .06 or .10. In these cases, a researcher might say 'the findings approached significance.' However, the American Psychological Association publication manual officially discourages reports of trends. The manual claims that if results do not meet the .05 level of significance, then they are to be interpreted as chance findings.

The decision to reject or retain the null hypothesis has been called dichotomous significance testing. Apparently the need for dichotomous significance testing grew out of the early history of statistics which developed in agriculture. Many of the statistical questions that an agriculturist might ask would be dichotomous in nature; for example, is the manure effective? It is easy to see in this example how a yes or no answer is appropriate and practical. However, it has been noted, particularly in psychology, that dichotomous significance testing has no clear or early theoretical basis. Thus, two contemporary theoreticians (Rosnow & Rosenthal, 1989) have said:

> 'that surely, God loves the .06 nearly as much as the .05. Can there be any doubt that God views the strength of evidence for or against the null as a fairly continuous function of the magnitude of p?'

DIRECTIONAL OR NONDIRECTIONAL ALTERNATIVE HYPOTHESES: ADVANTAGES AND DISADVANTAGES

Most statisticians use nondirectional alternative hypotheses. The advantage of the nondirectional alternative hypothesis is that it is less sensitive to chance differences in the data. Thus, the null hypothesis is less likely to be rejected and a Type One Error is less likely to be committed if a nondirectional alternative hypothesis is used. Since the Type One Error is considered more serious than the Type Two Error, the nondirectional alternative hypothesis is more attractive to statisticians.

However, the nondirectional alternative hypothesis has one major disadvantage, and that is, it is less sensitive to real differences in the data compared to the directional alternative hypothesis. For example, in testing to see whether two groups' means are significantly different from each other, if there is a real difference, but not a great difference, the nondirectional alternative hypothesis is less sensitive to this small but real difference between means.

Thus, it also follows that the directional alternative hypothesis has the advantage that it is more sensitive to real differences in the data. In other words, if there is a real difference between two groups' means, it is more likely to be detected with a directional alternative hypothesis. However, its major disadvantage is that it is also more sensitive to chance differences between two groups' means.

DID NUCLEAR FUSION OCCUR?

Two chemists claimed that they produced nuclear fusion in a laboratory under 'cold' conditions; that is, they claimed to have produced a vast amount of energy by fusing atoms and without having to provide large amounts of energy in order to do so. Their claims can still be analyzed in the hypothesis testing situation, although it is still not known conclusively whether they did or did not produce fusion.

The null and alternative hypotheses in this situation:

H_o: Fusion has not been produced.

H_a: Fusion has been produced.

Situation 1. If subsequent research supports their claims, then they made the correct decision to reject H_o. Thus, they will probably receive the Nobel Prize, and their names will be immortalized.

Situation 2. If subsequent research shows that they did not really produce fusion, then they rejected H_o when H_o was true, and thus, they committed the grievous Type One Error. Why is this a serious error? They may have misled thousands of researchers and millions of dollars may have been wasted. The money and resources might have been better spent pursuing their original lines of research rather than their mistake.

What about the quiet researcher who actually did demonstrate a small but real amount of fusion in the laboratory, but used a nondirectional alternative hypothesis? The researcher failed to reject H_o when H_a was true, and thus, the researcher committed a Type Two Error. What was the researcher's name? We do not know. Fame will elude a researcher if there is a continual commission of Type Two Errors because of an inordinate fear of Type One Errors! Remember, sometimes scientists must dare to be wrong.

HISTORY TRIVIA

The Karl Pearson Connection

Egon S. Pearson (1895–1980) was the son of the famous statistician Karl Pearson. Egon was born and reared in England, and he studied under his father at University College in London in the department of applied statistics. In 1922, Egon began to publish articles which would establish him as an important theoretical figure in modern statistics. In the 1920s, Egon Pearson also began an important collaboration with Jerzy Neyman (1894–1981) who would help shape one of Egon's most important contributions, statistical hypothesis testing.

Neyman grew up in Ukraine. He studied mathematics early in his college training, and he moved to Poland in 1921 and lectured in mathematics and statistics. He received his doctorate in 1924 at the University of Warsaw. In 1925, he received a fellowship to study statistics in England at the University College with Karl Pearson. Here Neyman would meet W. S. Gosset who would ultimately make his own unique contributions to statistics – the development of statistical analysis with small samples. Gosset introduced Neyman to R. A. Fisher (who among his other contributions developed the analysis of variance). Neyman was also to meet Egon S. Pearson who worked as an assistant in his father's statistics laboratory.

Egon S. Pearson, with the mathematical help of Neyman, further developed the notion of hypothesis testing including the testing of a simple hypothesis against an alternative, developed the idea of two kinds of errors in hypothesis testing, and proposed the likelihood ratio criterion which can be used to choose between two alternative hypotheses by comparing probabilities. With respect to the interactions between Gosset, Egon Pearson, and Neyman, it has been said that Gosset asked the question, Egon Pearson put the question into statistical language, and Neyman solved it mathematically.

In 1933, when Karl Pearson retired, the applied statistics department was split in two. R. A. Fisher became the Galton Professor succeeding Karl Pearson. Egon Pearson was appointed reader and, later, professor of statistics. In 1936, Egon took over the editorship of the famous journal, *Biometrika*, which was founded by his father to study mathematical and statistical contributions to life sciences. He remained editor for 30 years until his own retirement in 1966.

According to Helen Walker, a statistician and statistics historian, the modern history of statistics could be summarized thus. The first great wave of theoretical contributions came from Francis Galton and Karl Pearson. They promoted the idea that statistical analysis would provide important information, heretofore unknown, about people, plants, and animals. With their far reaching direction and influence even medicine and society would be positively changed by the science of statistics. Their contributions also included the invention of measures of association like the correlation coefficient and chi-square analysis, and the construction and publication of tables of statistics which were needed by statisticians and biometricians.

The second wave was begun by W. S. Gosset and completed by R. A. Fisher. According to Walker, this period was characterized by the development of statistical methods with small samples, initial development of hypothesis testing and design of experiments, and the development of criteria to choose among statistical tests.

The third wave was led by Egon S. Pearson and Jerzy Neyman. During their 10 year collaboration, the science of statistics enjoyed an ever-increasing popularity and appreciation. Hypothesis testing was refined and the logic of statistical inference was developed. The notion of confidence intervals was created, and ideas for dealing with small samples were further advanced and refined.

Although Karl Pearson held some controversial views, his contributions to the science of statistics are tremendous. His idea that the scientific method and statistical analysis should be considered the 'grammar of science' is a watershed in the history of statistics, and his title 'the founder of the science of statistics' is well-deserved.

KEY SYMBOLS AND TERMS

Archetypal Experiment – a two group experiment, with one group designated as the experimental group and one as the control group. The parameter of statistical interest is the difference between the two groups' means.

Signal-to-Noise Ratio – borrowed from signal detection theory where the effect of a treatment is considered the signal, and random variation in the numbers is considered the noise.

Research Hypothesis – also called the alternative hypothesis. It is most frequently what the experimenters think may be true, or wishes to be true before they begin an experiment. It can also be considered the experimenter's hunch.

Null Hypothesis – is the starting point in scientific research where the experimenter assumes there is no effect of the treatment or no relationship between two variables.

Alternative Hypothesis – also called the research hypothesis. It is most frequently what the experimenters think may be true, or wishes to be true before they begin an experiment. It can also be considered the experimenter's hunch.

Nondirectional Alternative Hypothesis – also called a two-tailed test of significance where the null hypothesis will be rejected if either Group

1's mean exceeds Group 2's mean, or vice versa, or where the null hypothesis will be rejected if a relationship exists, regardless of its nature.

Directional Alternative Hypothesis – also called a one-tailed test of significance where the alternative hypothesis is specifically stated beforehand. For example, Group 1's mean is greater than Group 2's mean.

Replication – refers to a series of experiments after an initial study where the series of experiments varies from the initial study in types of subjects, experimental conditions, etc. Replication should be conducted not only by the initial study's author but also other scientists who do not have a conflict of interest with the eventual outcome.

Type One Error – also known as alpha, and occurs when an experimenter rejects the null hypothesis when it is true.

Type Two Error – also known as beta, and occurs when an experimenter retains the null hypothesis when it is false.

Alpha – is the probability of committing the Type One Error. In order to consider findings significant, the probability of alpha must be less than .05.

Beta – is the probability of committing the Type Two Error. A Type Two Error can occur only when the null hypothesis is false and the experimenter fails to reject the null hypothesis.

Significance – findings are considered statistically significant if the probability that we are wrong (where we reject H_o and H_o is true) is less than .05. Significant findings indicate that the results of the experiment are substantial and not due to chance.

Nonsignificance – findings are considered statistically nonsignificant if the probability that we are wrong is greater than .05. Nonsignificant findings indicate that the null hypothesis has been retained, and the results of the experiment are attributed to chance.

Insignificant – is a value judgment, like deciding between good and evil, worthless and valuable. It typically has no place in statistics.

Trend – frequently reported when the data does not reach the conventional level of statistical significance (.05), but comes close (like .06 or .07). The American Psychological Publication Manual officially discourages reports of trends.

p **Level** – is the probability of committing the Type One Error; that is, rejecting H_o when H_o is true.

1. State H_o and H_a for the following problems:

   a. the relationship between the drug AZT and the AIDS virus.

   b. the relationship between mercury and multiple sclerosis.

   c. the relationship between aluminum and Alzheimer's disease.

   d. the relationship between copper bracelets and arthritis.

   e. the relationship between vitamin C and the flu.

   f. the relationship between thalidomide and birth defects.

2. Label the following decisions:

   a. Drug x really does reduce anxiety and the study retains H_o.

   b. Drug x really does reduce anxiety and the study rejects H_o.

   c. Drug x really does not reduce anxiety and the study retains H_o.

   d. Drug x really does not reduce anxiety and the study rejects H_o.

3. After a study rejects H_o, what is the next most important step in the scientific method?

TRUE/FALSE

4. A Type Two Error is considered less serious than a Type One Error.

5. The probability of making a Type One Error is referred to as alpha.

6. In comparison to a directional alternative hypothesis, a nondirectional alternative hypothesis is more sensitive to real differences in data.

7. The null hypothesis always represents the conservative position.

8. The conventional level of significance is $p < .01$.

9. An experimenter should report the lowest alpha level possible.

10. God loves the .06 level as much as the .05.

11. When a scientist replicates an experiment with identical results, then it is said the results are proven.

12. If results indicate nothing has happened, we would say they are insignificant.

13. Most statisticians use the nondirectional alternative hypothesis.

There is no statistical technique more useful nor more abused than correlation. **Correlation** is a statistical method which can determine the degree of relationship between two variables. Relationships between two variables can vary from strong to weak. When a relationship is strong, this means that knowing a person's score on one variable helps to predict their score on the second variable (like whether they will score high or low). If the relationship is a weak one, then knowing a person's score on one variable does not help to predict their score on the second variable.

The correlation value (or coefficient) ranges from +1.00 to –1.00.

```
|- - - - - - - - - - -|- - - - - - - - - - -|- - - - - - - - - - -|- - - - - - - - - -|
```

–1.00	–.50	0.00	+.50	+1.00

strong negative relationship   weak or none   strong positive relationship

When the correlation coefficient approaches +1.00 (or greater than +.50) it means there is a **strong positive relationship** or high degree of relationship between the two variables. This also means that the higher the score of a participant on one variable, the higher the score will be on the other variable. Also, if a participant scores very low on one variable then the score will be low on the other variable. For example, there might be a positive correlation between the number of calories a person eats and the individual's total weight. Therefore, the higher the number of calories eaten, the higher the total weight, and the lower the number of calories eaten, the lower the weight.

When the correlation coefficient approaches –1.00 (or less than –.50), it means that there is a **strong negative relationship**. This means that the higher the score of a participant on one variable, the lower the score will be on the other variable. For example, there might be a strong negative relationship between the average number of miles run per day and a person's resting blood pressure: the higher the number of miles run per day, the lower the blood pressure; the lower the number of miles run per day, the higher the blood pressure.

A correlation coefficient which is close to 0 (zero) means that knowing a person's score on one variable tells you nothing about their score on the other variable. For example, there might be a zero correlation between the number of letters in a person's last name and the number of miles driven per day. If you know the number of letters in a last name, it tells you nothing about how many miles they drive per day. There is no relationship between the two variables; therefore, there is a zero correlation.

The correlational statistical technique is used with correlational designs. In a correlational design, the experimenter has little or no control over the variables to be studied. The variables may be statistically analyzed long after they were initially produced or measured. Such data is called **archival**. The experimenter no longer has any experimental power to control the gathering of the data. It has already been gathered, and the experimenter now has only statistical power in his or her control. Cronbach (1967), a contemporary statistician, stated well the difference between the experimental and correlational techniques, '. . . the experimentalist [is] an expert puppeteer, able to keep untangled the strands to half-a-dozen independent variables. The correlational psychologist is a mere observer of a play where Nature pulls a thousand strings.'

One of the potential benefits of a correlational analysis is that *sometimes* a strong correlation between two variables may provide clues about possible cause-effect relationships. Correlational techniques applied to a large number of variables may allow a researcher to develop ideas about potential cause-effect relationships between variables. For this and other reasons, correlational designs and correlational statistical analyses are probably the most ubiquitous in all of science. However, it is not just their frequency that contributes to their continued abuse, it is also their very nature.

CORRELATION: USE AND ABUSE

The crux of the nature and the problem with correlation is that, just because two variables are correlated, it does not mean that one variable *caused* the other. We mentioned earlier a governor who wishes to supply every parent of a newborn child in his state with a classical CD or tape in order to boost the child's IQ. The governor supported his decision by citing studies which have shown a positive relationship between listening to classical music and intelligence. The governor is making at least two false assumptions. First, he is assuming a **causal relationship** between classical music and intelligence; that is, classical music causes intelligence to rise. However, the technique of correlation does not imply causation. And if x and y are correlated, then x is related to y, and y is related to x. It may not be that classical music increases intelligence (x is related to y), but maybe more intelligent people listen to classical music (y is related to x). Correlation does not distinguish nor give us any guidance *whatsoever* about when x is correlated with y whether it is because x is related to y or whether y is related to x. Second, the governor is making the mistake not only of basing his decision on a few correlational studies (when he should wait for evidence from experiments) but also he has not waited for scientific replication. It is far too early to assume that classical music raises people's intelligence based on a few correlational studies. Science must be cautious. These findings must be replicated through experiments in a variety of settings with a variety of people.

A WARNING: CORRELATION DOES NOT IMPLY CAUSATION

A major caution must be reiterated. Correlation does not imply causation. Because there is a strong positive or strong negative correlation between two variables, this *does not* mean that one variable is caused by the other variable. Many statisticians claim that a strong correlation *never* implies a cause-effect relationship between two variables. There are daily published abuses of the correlational design and statistical technique. A sampling of these misinterpretations follows:

1. Marijuana use and heroin use are positively correlated. Some drug opponents note that heroin use is frequently correlated with marijuana use. Therefore, they reason that stopping marijuana use will stop the use of heroin. Clear-thinking statisticians note that even a higher correlation is obtained between the drinking of milk in childhood and later adult heroin use. Thus, it is just as absurd to think that if early milk use is banned, subsequent heroin use will be curbed, as it is to suppose that banning marijuana will stop heroin abuse.

2. Milk use is positively correlated to cancer rates. While this is not a popular finding within the milk industry, there is a moderately positive correlation with drinking milk and getting cancer (Paulos, 1990). Could drinking milk cause cancer? Probably not. However, milk consumption is greater in wealthier countries. In wealthier countries people live longer. Greater longevity means people live long enough to eventually get some type of cancer. Thus, milk and cancer are correlated but drinking milk does not cause cancer (nor does getting cancer cause one to drink more milk).

3. Weekly church attendance is negatively correlated with drug abuse. A recent study demonstrated that adolescents who attended church weekly were much less likely to abuse illegal drugs or alcohol. Does weekly church attendance cause a decrease in drug abuse? If the Federal Government passed a law for mandatory church attendance, would there be a decrease in drug abuse? Probably not, and there might even be an increase in drug abuse. This study is another example of the abuse of the interpretation of the correlational design and statistical analysis.

4. Lead levels are positively correlated to antisocial behavior. A 1996 correlational study (Needleman *et al.*) examined the relationship of lead levels in the bones of 301 children and found that higher levels of lead were associated with higher rates of antisocial behavior. 'This is the first rigorous study to demonstrate a significant association between lead and antisocial behavior,' said one environmental health professor about the study. While the study's authors may have been very excited, the study is still correlational in design and analysis, thus, implications of causation should have been avoided. Perhaps antisocial children have a unique metabolic chemistry such that their bodies do not metabolize lead like normal children. Perhaps lead is not a *cause* of antisocial behavior but the *result* of being antisocial. Therefore, the reduction of lead exposure in early childhood may not reduce antisocial behavior at all. Also, note that there was a statistically 'significant' relationship. As you will learn later in this chapter, with large samples (like 301 children in this study) even very weak relationships can be statistically significant with correlational techniques.

5. The risk of getting Alzheimer's Dementia is negatively correlated with smoking cigarettes. In studies funded by the tobacco industry, it was found that the risk of getting Alzheimer's Dementia was negatively correlated with smoking cigarettes (yes, the risk went *down* with an increased use of cigarettes!). The implication of these findings for the tobacco industry was that increases in smoking (probably from increases in nicotine) would prevent the onset of Alzheimer's Dementia. This serves as a good example of the error in assuming a causal relationship because a correlation exists between two variables, and a demonstration of how a third variable may be controlling the other two variables. The risk of getting Alzheimer's Dementia increases with longevity. While about 10% of people over 65 are diagnosed with Alzheimer's, nearly 50% of those over age 90 are diagnosed with Alzheimer's. Heavy smokers die at a rate of about 500,000 a year. Smokers do not get the chance to get Alzheimer's Dementia because they do not live long enough.

6. Sexual activity is negatively correlated with increases in education. A 1997 report based on 10,000 interviews found that those with less education reported more sexual contacts per year than those who had been to post-graduate schools (Ph.D. programs, law school, medical school, etc.). Again, this is another example of a correlational design with a correlational analysis, therefore, causation cannot be inferred in these findings. And, once again, age may be the mitigating factor in this study. People with less than a baccalaureate degree tend to be younger than those with post-graduate degrees. Younger people tend to be more sexually active than older people. The study did not control for the age of the participants.

7. An active sex life is positively correlated with longevity. The 1997 newspaper headline for this study published in a British medical journal read, 'Study suggests frequent sex equals long life.' The study was conducted on a sample of 918 Welsh men divided into three groups: sex twice or more a week, an intermediate group, and those who had sex less than once a month. In a ten-year follow-up, the sexually inactive group had the highest death rate. The authors said the results could not be attributed to age or health. The authors, however, did not control for psychological variables like depression which could have been the precursor of physical disease. Thus, depression may have lowered sexual interest and subsequent physical disease may have accounted for the increase in death rates. Imagine doctors prescribing or ordering their patients to have frequent sex in order to increase longevity. It is quite possible we might witness a sudden increase in death rates. In order to change this correlational design to an experiment, the authors should have randomly assigned the men to one of the three sexual frequency groups, ordered the men to have sex according to the group they were assigned, and then assessed their longevity ten years later. Of course, this experimental design is not feasible but any causal inferences from the original correlational design are equally unfeasible.

8. Coffee drinking is negatively correlated with suicidal risk. In this 1996 study published in an American medical journal, 86,626 registered nurses were evaluated over a 10 year period. Increased coffee drinking was associated with lower rates of suicide. The authors concluded that the caffeine in the coffee might enhance mood and well-being resulting in lowered suicide risk. While the number of nurses studied is impressive, once again this study is a correlational design, and causation cannot be inferred. A stronger argument might have been made for the positive mood hypothesis had the authors randomly assigned depressed patients to groups, then prescribed various levels of caffeine. If the groups receiving the highest levels of caffeine had a subsequently higher sense of well-being and lowered suicide rates, then the hypothesis might be more plausible.

ANOTHER WARNING: CHANCE IS LUMPY

The contemporary Yale statistician Robert Abelson (1995) postulated Abelson's First Law of Statistics: **Chance is lumpy.** By this he means that if 86,626 nurses are measured for hundreds of variables over a ten year period, it would be highly surprising if dozens of significant relationships were *not* found! Or as my brother would say as he defends another one of his get-rich-quick schemes, 'Throw enough mud up on a wall, and some of it's bound to stick.' What Abelson is concerned with is that people fail to appreciate that long runs of occurrences can often be attributed to pure chance or random processes. Often people will attribute some unusual finding or run of luck to a mysterious or a systematic process when, in fact, only chance is operating. As Abelson notes, 'attributing a data set to mere chance is deflating.'

Abelson also notes that there is the psychological tendency to minimize the great variability that exists in small samples. Thus, we tend to overestimate the generalizability of small samples, when in reality, our conclusions may have varied widely across other samples.

There is no clear cut solution to the chance-is-lumpy problem. However, we should keep in mind that scientists should be somewhat conservative in their conclusions, be mindful of the repercussions of their findings and equally mindful of the tricks that can occur when we measure hundreds of variables in huge samples or a few variables in small samples.

CORRELATION AND PREDICTION

The correlation coefficient may also be used as an indicator of prediction. If a strong positive or negative correlation is obtained, then the relationship between the two variables may be likened to a predictive relationship. For example, in the previous strong negative relationship that was suspected between miles run per day and resting blood pressure, it could be said that the number of miles per day helps to predict resting blood pressure. If in reality a strong negative correlation exists, then a higher number of miles run per day predicts a lower resting blood pressure. For this same sample of participants, the predictions could also be stated vice versa, that is, a lower resting blood pressure predicts a higher number of miles run per day, or a higher resting blood pressure predicts a lower number of miles run per day.

If there was no relationship between the variables, or the correlation coefficient is close to or equal to zero, then it could be said that knowing a participant's score on one variable predicts nothing about the participant's score on the other variable.

THE FOUR COMMON TYPES OF CORRELATION

1. **Pearson's r:** A measure of the strength of a relationship between two continuous variables.

2. **Spearman's r:** A measure of the similarity between two ordinal rankings of a single set of data.

3. **Point-Biserial r:** A measure of the strength of a relationship between one continuous variable and one dichotomous variable (a two-level-only variable like gender).

4. **Phi (ϕ) Correlation:** A measure of the strength of a relationship between two dichotomous variables.

THE PEARSON PRODUCT-MOMENT CORRELATION COEFFICIENT

The single most common type of correlation is the **Pearson Product-Moment Correlation Coefficient** which measures the degree of relationship between two continuous variables. A continuous variable is a variable which can be measured along a line scale. For example, IQ values are continuous because they can range along a line scale from 0 to about 200. Gender (male or female) is not considered a continuous variable because if numbers (like 1 or 2) were assigned to the two categories, a person could not be a 1.3 or a 1.7. The story of the naming of the correlation coefficient appears at the end of this chapter in the History Trivia section.

Pearson's coefficient r is obtained for a sample drawn from a population. The population value of Pearson's coefficient is called **rho** (ρ), and thus, r is an estimate of p.

The formula for r is as follows:

$$r = \frac{N\Sigma xy - (\Sigma x)(\Sigma y)}{\sqrt{[N\Sigma x^2 - (\Sigma x)^2][N\Sigma y^2 - (\Sigma y)^2]}}$$

Note: In this formula N is equal to the number of pairs of scores and Σxy is called the sum of the cross products. Let's see how the formula works in the following example.

An experimenter wishes to know whether heavy smoking is related to longevity. From a sample of recently deceased smokers, the number of cigarettes (estimated per day for their last five years after visits with their surviving relatives) is paired with the number of years that they lived.

Subject	Cigarettes	Years lived
1	25	63
2	35	68
3	10	72
4	40	62
5	85	65
6	75	46
7	60	51
8	45	60
9	50	55

We will arbitrarily name one variable x, and the other variable y. The results of Pearson's r will be exactly the same no matter which variable is labeled x or y.

Step 1. First, obtain Σx, Σx^2, $(\Sigma x)^2$, and $(\Sigma y)^2$

$\Sigma x = 25 + 35 + 10 + 40 + 85 + 75 + 60 + 45 + 50 = 425$

$\Sigma x^2 = 25^2 + 35^2 + 10^2 + 40^2 + 85^2 + 75^2 + 60^2 + 45^2 + 50^2 = 24,525$

$\Sigma y = 63 + 68 + 72 + 62 + 65 + 46 + 51 + 60 + 55 = 518$

$\Sigma y^2 = 63^2 + 68^2 + 72^2 + 62^2 + 65^2 + 46^2 + 51^2 + 60^2 + 55^2 = 33,188$

$(\Sigma x)^2 =$ HINT: square Σx or $(25 + 35 + 10 + 40 + 85 + 75 + 60 + 45 + 50)^2$

$(\Sigma x)^2 = 425^2 = 180,625$

$(\Sigma y)^2$ = HINT: square Σy or $(63 + 68 + 72 + 62 + 65 + 46 + 51 + 60 + 55)^2$

$(\Sigma y)^2 = 542^2 = 293,764$

Step 2. Obtain the sum of the cross products (Σxy) by multiplying each x score by its paired y score.

$\Sigma xy = (25 \times 63) + (35 \times 68) + (10 \times 72) + (40 \times 62) + (85 \times 65) + (75 \times 46) + (60 \times 51) + (45 \times 60) + (50 \times 55)$

$\Sigma xy = 1575 + 2380 + \ldots + 2750$

$\Sigma xy = 24,640$

Step 3. Obtain the value of the numerator in the r formula. Remember N is equal to the number of pairs of scores. In this example there are 9 pairs of scores.

$N\Sigma xy - (\Sigma x)(\Sigma y)$

$(9)(24,640) - (425)(542)$

$221,760 - 230,350 = -8,590$

Step 4. Obtain the value of the denominator in the r formula. Remember that the square root is obtained after all other denominator values have been computed, simplified, and reduced to a single number.

$$= \frac{}{\sqrt{[N\Sigma x^2 - (\Sigma x)^2][N\Sigma y^2 - (\Sigma y)^2]}}$$

$$= \frac{}{\sqrt{[(9)(24,525) - (425)][(9)(33,188) - (542)]}}$$

$$= \frac{}{\sqrt{[220,725 - 180,625][298,692 - 293,764]}}$$

$$= \frac{}{\sqrt{[40,100][4,928]}}$$

$$= 14,057.482$$

Step 5. Divide the result of Step 3 by the result of Step 4.

$r = \dfrac{-8,590}{14,057.482}$

$r = -0.6111$

$r = -0.61$

Note that the Pearson's r is usually rounded off to two decimal places. Thus, r = –0.61 means that there is a strong negative correlation between smoking and longevity. This indicates that the higher the number of cigarettes smoked in the past five years, the lower the number of years lived. And the lower the number of cigarettes, the higher the number of years lived. Remember, this relationship between these two variables DOES NOT mean that heavy smoking causes one to live fewer years. It may, however, give clues as to further research ideas for experiments. In this case, an experiment might be set up (perhaps with animals) with an experimental group and a control group to determine whether cigarette smoking actually has a causal relationship with early morbidity.

TESTING FOR THE SIGNIFICANCE OF A CORRELATION COEFFICIENT

A correlation coefficient may be tested to determine whether the coefficient significantly differs from zero. The value r is obtained on a sample. The value rho (ρ) is the population's correlation coefficient. It is hoped that r closely approximates rho. The null and alternative hypotheses are as follows:

H_o: $\rho = 0$

H_a: $\rho \neq 0$

The value of r and the number of pairs of scores are converted through a formula into a distribution (similar to the z distribution) called the **t distribution** (in Appendix B). The t formula can only be used to test whether r is equal to zero. It cannot be used to test to see whether r might be equal to some number other than zero. It is also important to note that the t distribution may be used to test other types of inferential statistics. Therefore, if someone says that a t test is being used, it would be a legitimate question to ask why. The t distribution is most commonly used to test whether two means are significantly different; however, it may also be used to test the significance of the correlation coefficient. The t distribution also has other uses. Interestingly, the t distribution becomes the z distribution when the data is infinite, but they are also strikingly visually similar when there are only several hundred numbers in the set of data.

The t test formula in order to test the null hypothesis for a correlation coefficient is:

$$t = \frac{r}{\sqrt{\dfrac{1 - r^2}{N - 2}}}$$

In the previous example on smoking, the research question was whether heavy smoking was related to longevity. A $r = -.61$ was obtained which meant that there was a strong negative relationship between smoking and longevity. In order to test whether this obtained r significantly differs from zero, the t formula is used.

$$t = \frac{r}{\sqrt{\dfrac{1 - r^2}{N - 2}}}$$

where   N = the number of pairs of scores.

$$t = \frac{-0.6111}{\sqrt{\dfrac{1 - (-0.6111)^2}{9 - 2}}}$$

$$t = \frac{-0.6111}{\sqrt{\dfrac{1 - 0.3734}{7}}}$$

$$t = \frac{-0.6111}{\sqrt{\dfrac{0.6266}{7}}}$$

$$t = \frac{-0.6111}{\sqrt{0.0895}}$$

$$t = \frac{-0.6111}{0.2992}$$

$$t = -2.042$$

OBTAINING THE CRITICAL VALUES OF THE *t* DISTRIBUTION

We will now obtain the **critical values** of *t* which is obtained from the *t* distribution in Appendix B.

Step 1. Choose a one-tailed or two-tailed test of significance. The alternative hypothesis establishes whether we will use a one-tailed or two-tailed significance test. If the alternative hypothesis is non-directional, as is the case in most studies, then a two-tailed test of significance is required.

Step 2. Choose the level of significance. The conventional level of significance is $p = .05$. Only in rare circumstances would one ever depart from $p = .05$ as a starting point.

Step 3. Determine the degrees of freedom (*df*). The *df* is an advanced statistical concept related to sampling. We will keep things simple: The formula for this *t* test statistic is $df = N - 2$ where N is the number of pairs of scores. In this example, there were 9 pairs of scores so $df = 9 - 2$ or $df = 7$.

Step 4. Determine whether the *t* from the formula (called the derived *t*) exceeds the tabled critical values from the *t* distribution. For a two-tailed test of significance at $p = .05$ with $df = 7$, the critical values of *t* are $t = +2.365$ and $t = -2.365$. If the derived *t* is greater than $t = +2.365$ or less than $t = -2.365$, then the null hypothesis will be rejected. In this example, the derived $t = -2.042$ is not less than $t = -2.365$, therefore, the null hypothesis is not rejected, and it will be concluded that $r = -.61$ indicates a nonsignificant relationship. Curiously, although the strength of the relationship was strong ($r = -.61$), the test of significance indicated that the obtained relationship was likely due to chance or there was greater than 5 chances out of 100 that the relationship was due to chance.

In a research paper, the results might be reported as follows:

'There was a strong negative correlation found between smoking and longevity although the correlation was not statistically significant, $r(7) = -.67, p > .05$.' Note that the degrees of freedom appear in the parentheses to the right of *r*.

If the Null Hypothesis is Rejected

Remember also that if the null hypothesis is rejected, the experimenter would report the lowest *p* level possible. In that case, the derived *t* would be compared to the critical values of *t* at .01 and .001. If the derived *t* exceeded the critical values at both .01 and .001, then the experimenter would report the *r* as significant at $p < .001$.

REPRESENTING THE PEARSON CORRELATION GRAPHICALLY: THE SCATTERPLOT

A **scatterplot** is a graphic presentation of the pairs of scores involved in a correlation coefficient. It is also sometimes called a **bivariate distribution**. Let us produce a scatterplot of the example on page 116 of the number of cigarettes smoked and longevity. In order to construct a scatterplot, prepare to graph each of the variables along one of the graph's axes. Ultimately, it does not matter which variable is chosen for which axis, just as long as you prepare each axis for one of the variables (see Figure 6.1).

The order in which the pairs are graphed is not important. Thus, it does not matter if you begin with the last pair of scores or the first pair of scores. However, each *pair* of scores is very important, so be sure to plot each participant's score on one variable with their *corresponding score* on the other variable. For example, if we began with the participant who smoked an average of 85 cigarettes per day and lived 65 years, we would first locate the 85 cigarettes value on the cigarette axis and draw a horizontal line across the graph along this value. Next, locate this participant's longevity score on the longevity axis and draw a line vertically up the graph. The intersection of the two lines is the graphic representation of the pair of values in mathematical two-dimensional space. Figure 6.2 shows a completed scatterplot.

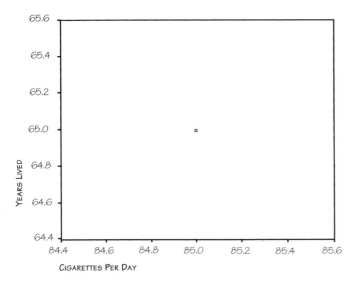

Figure 6.1

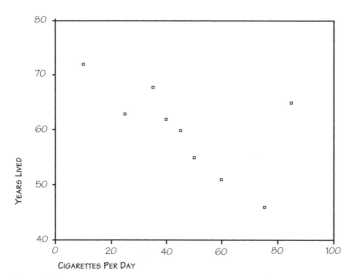

Figure 6.2

FITTING THE POINTS WITH A STRAIGHT LINE: THE ASSUMPTION OF A LINEAR RELATIONSHIP

When using the Pearson correlation coefficient, it is assumed that the cluster of points is best fit by a straight line. Look at the cluster in Figure 6.2, and imagine how a straight line would pass through the points with as little overall distance between all the points and the straight line as possible. Obviously there is no single straight line that would pass through all the points. There is only a best-fitting line that minimizes all of the distances between the points and the line. If a straight line, as opposed to a curved line, best fits the points then the relationship between the two variables is said to be **linear**. If a curved line best fits the points, then the relationship between the two variables is said to be **curvilinear**. Remember it is an assumption of the correlation coefficient that the best fitting line is linear, or in other words, it is assumed that the relationship between the two variables is linear. The violation of this assumption is typically not harmful, at least in terms of committing a Type One Error. If we assume a relationship is linear when it is really curvilinear, it will result in a lower r value, and statistical significance is less likely to be attained. Of course, it could conceivably be harmful in subtle ways to the experimenter if no relationship was found where one actually exists (Type Two Error). Figure 6.3 presents a curvilinear relationship. Curvilinear relationships are not uncommon. For example, there is a curvilinear relationship between strength and age: When we are younger we are weaker, when we are older we are stronger, but when we become very old, we are weaker again.

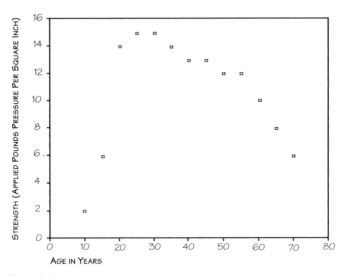

Figure 6.3

INTERPRETATION OF THE SLOPE OF THE BEST-FITTING LINE

The best-fitting line may also be used to interpret the nature and strength of the relationship between two variables. In Figure 6.2, the best-fitting line slopes from the upper left of the graph to the lower right. This indicates that there is a negative correlation between the two variables. The relatively small amount of overall distance between the points and the line indicates that this negative relationship is also strong; that is, it will be a large negative number.

A positive relationship will produce a best-fitting line that slopes from the lower left of the graph to the upper right. A weak or no relationship will produce a seemingly random cluster of points. There will be no best-fitting line, or it could be said that a straight horizontal line through the cluster is as good as any.

A perfect correlation ($r = 1.00$) would produce a scatterplot where the best-fitting straight line passes through all of the points, and thus, it is an extremely unlikely event in the real world of data. See Figure 6.4 for a graphic representation of positive, negative, weak, and perfect correlations.

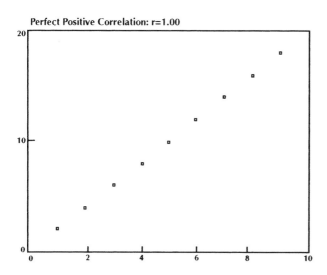

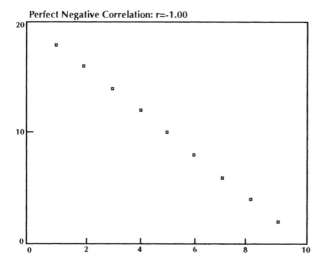

Figure 6.4 *(continued overleaf)*

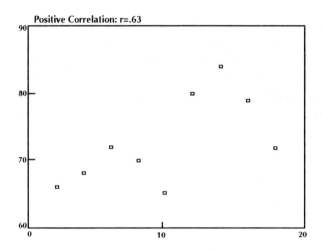

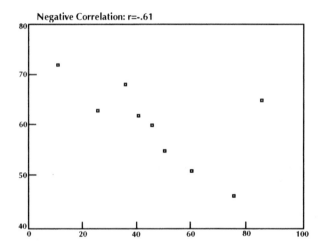

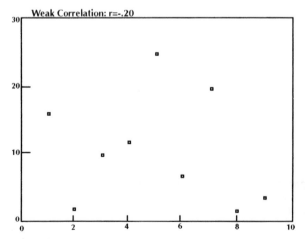

Figure 6.4 *(continued)*

THE ASSUMPTION OF HOMOSCEDASTICITY

A second assumption of the correlation coefficient is that of **homoscedasticity**. This assumption is met if the distance from the points to the line is relatively equal all along the line. The violation of the assumption is called **heteroscedasticity**, and a graphic representation is presented in Figure 6.5.

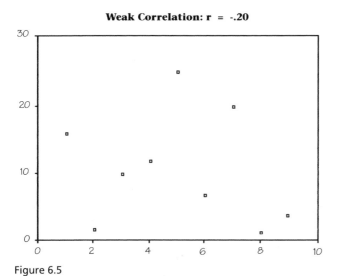

Weak Correlation: r = -.20

Figure 6.5

The effects of violating the assumption of homoscedasticity are the same as violating the assumption of linearity, and that is, the value of r is more likely to underestimate the population value of r.

THE COEFFICIENT OF DETERMINATION: HOW MUCH ONE VARIABLE ACCOUNTS FOR VARIATION IN ANOTHER VARIABLE: THE INTERPRETATION OF r^2

Another use of the correlation coefficient is its squared value. r^2 is called the **coefficient of determination**, and it has two important interpretations. First, it explains the proportion of variance in one variable accounted for by the other variable. For example, if the correlation between two variables A and B is $r = .25$, then $r^2 = (.25)(.25) = .0625$. Therefore, the variable A explains approximately 6% of the variation in variable B. Another way of looking at r^2 in this example, would be to say that 6% of the variation in variable B can be explained by its relationship to variable A, and 94% of the total variance between variables A and B remains unexplained. The former variance is called **shared variance** and the latter variance is called **uncommon, unshared,** or **unexplained variance**.

UNIVERSITY COLLEGE
Library

A second interpretation of r^2 is its use as a measure of strength between two or more rs. For example, if given an $r = .25$ and $r = .50$, it is clear that the second r value is twice as great as the first r value. However, the r^2 values are 6.25% and 25% respectively. Thus, $r = .50$ actually explains four times as much variance as does an $r = .25$. The moral here is to be careful when interpreting and comparing r values, particularly smaller values of r where any $r < .31$ will explain less than 10% of the variance in two variables.

The same causality caution that was applied to the interpretation of the simple correlation coefficient r is applied to the interpretation of r^2. The coefficient of determination does not provide an *explanation* for the observed relationship nor does it *account* for the *reason* for the relationship. The coefficient of determination simply provides another way of viewing the relationship between two variables.

QUIRKS IN THE INTERPRETATION OF SIGNIFICANT AND NON-SIGNIFICANT CORRELATION COEFFICIENTS

There are some serious quirks in the interpretations of the correlation coefficients. We have witnessed one already in our example on smoking and longevity. The interpretation of the *strength* of the relationship between these two variables was actually *independent* of the significance testing. The derived $r = -.61$ indicated that there was a moderately strong negative correlation between smoking and longevity. However, we found that this relationship was not statistically significant at $p < .05$. The statistical quirk is that significance tests are artifactually dependent upon sample size: Larger sample sizes will more likely produce significance than smaller samples. In our example, N = 9 was an exceptionally small sample, thus, despite a moderately strong r value, we were not able to reject the null hypothesis. If we had had 30 pairs of scores, we would have obtained significance.

At this point, the coefficient of determination helps us to interpret this sample size quirk. When we square $r = -.61$, we obtain .37 or 37% of the variance is shared by the two variables. This is not a small amount of variance, all things being equal. In a 1988 prejudice study, it was found that $r = .06$ was significant for the relationship between anti-Semitism and prejudice against smoking. How could the latter r be statistically significant when our $r = -.61$ was not? In the prejudice study there were 5,977 pairs of scores. When we square $r = .06$, we obtain $r^2 = .0036$ or about one-third of one percent of the total variance is explained variance! This also indicates that well over 99% of the variance remains unexplained between the two variables. In these two examples, the coefficient of determination has helped clear away some of the confusion in the interpretation of significance or a lack of significance.

SPEARMAN'S CORRELATION

Spearman's correlation coefficient (sometimes referred to as Spearman's rho or r_s where the sub s is in honor of Spearman) determines the degree of relationship for ranked data. Spearman's correlation is also called the rank-order correlation coefficient. Although Spearman's correlation is far less common than Pearson's r, occasionally variables are ordered according to rank (like 1st through 10th), or variables may be subsequently ranked on the basis of a continuous variable. The formula for Spearman's r:

$$r_s = 1 - \frac{6\Sigma D^2}{N(N^2 - 1)}$$

where N = the number of pairs of scores.

Spearman's might be most typically used in situations where there are a number of variables and they are all ranked by two independent judges. For example, a researcher wishes to know whether married couples have similar tastes in vegetables. The members of each couple were independently asked to rate their preference for seven vegetables from most preferred (#1 rank) to least preferred (#7). Their data is as follows:

	Husband	Wife	D score	D²
Broccoli	4	3	1	1
Cauliflower	3	1	2	4
Brussel Sprouts	6	7	−1	1
Okra	1	2	−1	1
Cabbage	5	5	0	0
Spinach	2	4	−2	4
Turnips	7	6	1	1

$$\Sigma D^2 = 12$$

Note that the D score is the difference between the pairs of ranks on the first variable ranked, etc. The number '6' in the formula is a **constant** and remains '6' regardless of the numbers of ranked variables. N is the number of pairs of ranks (or the number of variables that are ranked). In this example, N = 7.

$$r_s = 1 - \frac{6\Sigma D^2}{N(N^2 - 1)}$$

$$r_s = 1 - \frac{6(12)}{7(49 - 1)}$$

$$r_s = 1 - \frac{72}{7(48)}$$

$$r_s = 1 - \frac{72}{336}$$

$$r_s = 1 - 0.214$$

$$r_s = 0.786$$

$$r_s = 0.79$$

Note that Pearson's r and Spearman's r are most typically reported to two decimal places.

Spearman's r may be interpreted as a measure of the linear correlation between ranks. Pearson's r will produce the same value as Spearman's r on the same set of ranked data. In the case where the variables are expressed in their original form as continuous measures, the Pearson's r will not equal the Spearman's r after they have been converted to ranks, but they will have similar values.

SIGNIFICANCE TEST FOR SPEARMAN'S r

Spearman's r cannot be tested for significance in the same manner as Pearson's r. Refer to the significance table for Spearman's r in Appendix C. This table presents the minimum size of Spearman's r in order to reject the null hypothesis at $p < .05$ and $p < .01$. This table is unique because the degrees of freedom do not have to be calculated, since N is used directly in the table. In the previous example, the null and alternative hypotheses are:

$$H_o: r = 0$$

$$H_a: r \neq 0$$

The Spearman's significance table reveals that an r value of at least .786 is necessary to reject H_o at the $p < .05$ level. The obtained r_s value equals this level exactly, therefore, H_o can be rejected at $p = .05$. According to APA format the derived r_s value may be reported as:

$$r_s(7) = .79, p = .05$$

In conclusion, with respect to the previous example, the results indicate that there is a strong positive correlation between the husband's and the wife's preferences for the seven vegetables. The probability that r_s would equal .79 by chance alone is equal to five chances out of a hundred or .05. Thus, $r_s = .79$ may be reported as statistically significant.

TIES IN RANKS

The following example comes from a study (Coolidge, 1983) of Wechsler Intelligence Scale for Children-Revised (WISC-R) profiles of emotionally disturbed children (EDC) and learning disabled children (LDC). The WISC-R contains 10 separate subtests. The focus of the study was whether the relative strengths and weaknesses within each group of children were similar between the two groups. The subtests were ranked from highest (#1) to lowest (#10) in terms of overall group performance. The data is tabled as follows:

Subtest	EDC rank	LDC rank	D	D²
1	10	10	0	0
2	7	5.5*	1.5	2.25
3	9	9	0	0
4	8	8	0	0
5	3	1	2	4
6	6	7	−1	17
7	5	5.5*	−.5	.25
8	1	2	−1	1
9	2	4	−2	4
10	4	3	1	1
				$\Sigma D^2 = 13.5$

Note: Since there is a tie between two subtests at the 5th rank place, positions 5 and 6 are added together and divided by 2 for a 5.5 average rank for both subtests. An asterisk was added to indicate the tie. **Note that since the 5th and 6th ranked places are now taken, the next lowest subtest is given the 7th place rank.** Had a three way tie occurred at the 5th rank place, then places 5, 6, and 7 would have been added together and divided by three. Thus, all three tied subtests would be given a 6. The next subtest would be 8th ranked.

$$r_s = 1 - \frac{6\Sigma D^2}{N(N^2 - 1)}$$

$$r_s = 1 - \frac{6(13.5)}{10(100 - 1)}$$

$$r_s = 1 - \frac{81}{990}$$

$$r_s = 1 - 0.0818$$

$$r_s = 0.918$$

$$r_s = 0.92$$

The null and alternative hypotheses are:

$$H_o: \rho = 0$$

$$H_a: \rho \neq 0$$

According to the Spearman's significance table, H_o is rejected at $p < .01$. It may be concluded that there is a significant, strong positive correlation between the two sets of ranks, and the relative strengths and weaknesses within the groups are similar between the two groups. This means that knowing the rank of a sub-test in one group will predict the rank of the same subtest in the other group $r_s = .92$, $p < .01$.

POINT-BISERIAL CORRELATION

The point-biserial correlation (r_{pb}) gives an estimate of the degree of relationship between a dichotomous variable and a continuous variable. One typical use of r_{pb} is correlating a test question which is dichotomous with the overall test score which is continuous. This might tell the researcher whether an individual item is a good predictor of the overall test score. The formula is as follows:

$$r_{pb} = \frac{\bar{X}_1 - \bar{X}_2}{S} \cdot \sqrt{pq}$$

where

$\bar{X}_1 =$ the mean score on the continuous variable of just the participants on level one of the dichotomous variable

$\bar{X}_2 =$ the mean score on the continuous variable of just the participants on level two of the dichotomous variable

$S =$ the standard deviation of all the participants on the continuous variable

$p =$ the proportion of persons in level one of the dichotomous variable

$q = 1 - p$

For example, an experiment was performed to see if fine motor movement, as measured by speed of finger tapping, was related to gender. In this case gender is inherently a dichotomous variable while finger tapping (number of taps in 5 seconds) is measured as a continuous variable. The data is tabled as follows:

Subjects	Finger taps
Male	10
Female	12
Female	14
Male	9
Male	11
Female	13
Female	14
Female	10
Male	8
Female	11

Arbitrarily consider the males as level 1 and females as level 2 of the dichotomous variable.

Thus,

$\bar{x}_1$ = mean score on continuous variable of the level one group

$$= \frac{(10 + 9 + 11 + 8)}{4} = 9.50$$

$\bar{x}_2$ = mean score on continuous variable of the level two group

$$\bar{x}_2 = \frac{(12 + 14 + 13 + 14 + 10 + 11)}{6} = 12.33$$

S = standard deviation of continuous variable

$$S = \sqrt{\frac{\Sigma x^2 - \frac{(\Sigma x)^2}{N}}{N-1}}$$

$$S = \sqrt{\frac{1292 - \frac{(112)^2}{10}}{9}}$$

$S = 2.044$

p = proportion of level one to total

$$p = \frac{4}{10} = 0.4$$

$$q = 1 - p = 1 - 0.4 = 0.6$$

$$r_{pb} = \frac{9.50 - 12.33}{2.044} \cdot \sqrt{(0.4)(0.6)}$$

$r_{pb} = -0.679$

$r_{pb} = -0.68$

TESTING FOR THE SIGNIFICANCE OF THE POINT-BISERIAL CORRELATION COEFFICIENT

The significance of the point-biserial correlation coefficient is tested the same way as the Pearson's r. The null and alternative hypotheses are as follows:

$H_o: \rho = 0$

$H_a: \rho \neq 0$

The value of r and the number of pairs of scores are converted to a t distribution with $N - 2$ degrees of freedom where N is the number of pairs of scores. The t statistic can only be used to test whether r is equal to zero. The formula is as follows:

$$t = \frac{r}{\sqrt{\dfrac{1 - r^2}{N - 2}}}$$

where

N = the number of pairs of scores and $df = N - 2$.

Thus:

$$t = \frac{-0.6790}{\sqrt{\dfrac{1 - 0.4611}{8}}}$$

$$t = \frac{-0.6790}{\sqrt{0.0674}}$$

$$t = \frac{-0.6790}{0.2595}$$

$$t = -2.617$$

The critical value of t is obtained from the t distribution. In this case, the critical value of t with $df = 8$ and $p = .05$ is ± 2.306. Our formula-derived t of -2.617 exceeds the critical value of t, therefore, we reject H_o and conclude that our $r_{pb} = -.68$ is statistically significant. This means that there is a significant relationship between gender and finger tapping. Since males were level one, the interpretation of the negative correlation would be that gender level one (arbitrarily considered the higher [or first] level of the dichotomous variable) is associated with lower (or second) levels of the continuous variable, and the lower level of the dichotomous variable is associated with higher levels of the continuous variable.

In a research paper, the r_{pb} might be reported as follows:

'There was a strong negative correlation found between gender and fine motor movement $r_{pb}(8) = -.68, p < .05$.'

PHI ϕ CORRELATION

The phi (rhymes with fee) correlation gives an estimate of the degree of relationship between two dichotomous variables. The value of the phi ϕ correlation coefficient is interpreted just like the Pearson r, that is, it can vary from -1.00 to $+1.00$.

For example, a sample of 50 college students were given Beck's Depression Scale and the MMPI Depression Scale. The frequencies of the students who met the criterion for depression on each scale were noted and tabled as follows:

		MMPI Depression Scale	
		Depressed	Not Depressed
	Depressed	6 (a)	2 (b)
Beck's			
	Not Depressed	6 (c)	36 (d)

The individual cells in this matrix of numbers have been labeled 'a' through 'd' in order to identify the cells in the formula. The formula for phi is:

$$\text{phi} = \frac{ad - bc}{\sqrt{(a+b)(c+d)(a+c)(b+d)}}$$

$$\text{phi} = \frac{(6)(36) - (2)(6)}{\sqrt{(6+2)(6+36)(6+6)(2+36)}}$$

$$\text{phi} = \frac{216 - 12}{\sqrt{8 \times 42 \times 12 \times 38}}$$

$$\text{phi} = \frac{204}{\sqrt{153,216}}$$

$$\text{phi} = \frac{204}{391.4281}$$

$$\text{phi} = 0.52$$

TESTING FOR SIGNIFICANCE OF PHI

The phi correlation can be tested for significance by converting the value of phi into a chi-square statistic (χ^2) and comparing it to the **chi-square distribution**. The null and alternative hypotheses are as follows:

H_0: phi $= 0$

H_a: phi $\neq 0$

The formula for the conversion of phi to chi-square is:

$$\chi^2 = N \, (\text{phi})^2$$

where

> N = the number of participants in the correlation and the *df* is always equal to 1.

The critical values of chi-square are in Appendix D. The critical value of chi-square with $df = 1$ at $p = .05$ is 3.84. The obtained value of chi-square is $(50)(.52)^2 = .13.52$. The obtained value does exceed the critical value, therefore, H_o is rejected. It is concluded that there is a significant relationship between whether a person is scored depressed on Beck's scale and whether that person is scored depressed on the MMPI Depression Scale. In other words, knowing whether they are labeled as depressed on Beck's does predict whether they will be labeled depressed on the MMPI.

HISTORY TRIVIA

Francis Galton (1822–1911) is credited with the first formal presentation of a statistical relationship. He was profoundly influenced by Charles Darwin, and Galton's work was devoted to prediction as a tool for the study of inheritance. For part of his research, Galton used sweet pea seeds, and he ranked the size of the parent and offspring seeds. He also studied height in fathers and sons. From this research he noted that the offspring at their mature height showed less variability and fewer extremes than the parents. He called this phenomenon 'reversion' and thus was derived the symbol '*r*.'

Karl Pearson (1857–1936) was a friend of Galton, and he is likewise famous in the early history of statistics. In 1893, Pearson presented the term 'standard deviation.' In 1895, he published an article deriving the current correlation coefficient formula and its test statistic. In the 1920s, Pearson's son, Egon, developed the idea of hypothesis testing, and his work led to the present definitions of the null and alternative hypotheses.

In 1904, Charles Spearman (1863–1945) published an article in which he stated a formula for rank-order correlation. However, the present formula was actually derived by Karl Pearson in 1907. In addition, Galton was probably the first to develop the concept of correlation with ranks when he rank-ordered his sweet pea seeds. Furthermore, it was Pearson who proposed to use rho (ρ) as the symbol for the rank order correlation coefficient, although most statistics books use rho (ρ) as the symbol for the population correlation coefficient. r_s (where the s occurs in honor of Spearman) is also currently used as the rank-order correlation coefficient.

Correlation – measures the degree of relationship between two variables.

Strong Positive Relationship – indicates that a high score on variable x will be associated with a high score on variable y, and a low score on variable x will be associated with a low score on variable y.

Strong Negative Relationship – indicates that a high score on variable x will be associated with a low score on variable y, and a low score on variable x will be associated with a high score on variable y.

Archival Data – is a type of study which has already been conducted and the data has already been gathered. The experimenter only has statistical power and no longer has any means of changing the original experimental design.

Causal Relationship – should never be inferred from a correlational design or a correlational test statistic. Causal relationships can be inferred from true experimental designs (like the archetypal experiment).

Chance is Lumpy – is a rule of statistics where a long and unusual occurrence in data should first be attributed to randomness or chance.

Pearson Product Moment Correlation – is a measure of the strength of a relationship between two continuous variables. It is measured by the coefficient *r*.

Scatterplot – is a graphic representation of the relationship of two continuous variables in correlational designs. It is also called a bivariate distribution.

Linear Relationship – is an assumption made between two continuous variables when graphically represented. The assumption is that a straight line best fits the bivariate distribution.

Curvilinear Relationship – is an assumption made between two continuous variables when graphically represented. The assumption is that a curved line best fits the bivariate distribution. The Pearson's *r* cannot be used when the relationship is curvilinear.

Homoscedasticity – is an assumption of the Pearson Product Moment Correlation Coefficient that the bivariate distribution is evenly distributed throughout the length of the scatterplot best-fitting line.

Heteroscedasticity – is considered to be a violation of the assumption of homoscedasticity and is the condition where the bivariate distribution has greater variance on some lengths of the line than on others.

Coefficient of Determination (r^2) – is an interpretation made by squaring the Pearson's r and determines how much variance is accounted for in one variable by variation in another variable.

Spearman's Correlation – is a measure of the relationship between two ordinal rankings of the same set of data.

Point-Biserial Correlation – is a measure of the strength of a relationship between one continuous variable and one dichotomous variable.

Phi Correlation – is a measure of the strength of a relationship between two dichotomous variables.

Degrees of Freedom – is a complicated statistical concept, but is approximately and usually less than N of a set of numbers.

Chi-Square (χ^2) Distribution – is a distribution which can be used to test the significance of a Phi correlation and also to test nonparametric frequency data sets.

Constant – is a specific number that always stays the same in a statistical formula.

A researcher is interested in determining if there is a relationship between age and performance on the Block Design (BD) and Vocabulary (VOC) subtests of the WAIS-R. The results are as follows:

Age	VOC	BD	Age	VOC	BD
66	9	10	65	9	14
81	13	6	72	10	11
83	13	5	74	11	11
65	10	13	76	12	9
71	11	10	71	10	10
82	12	7	88	11	7
86	14	6	68	11	8
71	10	11	74	11	7
63	8	14	65	10	14
90	13	5	75	12	7

1. For the relationship between Age and Vocabulary

   a. Set up H_o and H_a.

   b. Produce a scatterplot.

   c. Perform an analysis using Pearson's r.

   d. Test for significance.

   e. Write up your results.

2. For the relationship between Age and Block Design

   a. Set up H_o and H_a.

   b. Produce a scatterplot.

   c. Perform an analysis using Pearson's r.

   d. Test for significance.

   e. Write up your results.

3. For the relationship between Vocabulary and Block Design

   a. Set up H_o and H_a.

   b. Produce a scatterplot.

   c. Perform an analysis using Pearson's r.

   d. Test for Significance.

   e. Write up your results.

4. A psychology professor wishes to know whether undergraduates and graduate students have similar areas of interest in psychology. Each student was asked to rate their preference for eight core areas in psychology (1 = most preferred, 7 = least preferred.)

	Undergraduate	Graduate
Statistics	1	3
Developmental	4	6
Assessment	1	2
Research	5	4
Cognitive	3	3
Learning	2	2
Abnormal	6	7
Social	3	1

a. Set up H_o and H_a.

b. Perform an analysis using Spearman's correlation.

c. Test for significance.

d. Write up the results.

5. An experiment was performed to see if ability on the Block Design subtest of the WAIS-R is related to gender. The results are as follows:

Gender	Time in seconds
Male	90
Female	92
Female	93
Male	91
Female	90
Male	89
Female	94
Male	89
Male	88

a. Set up H_o and H_a.

b. Perform an analysis using the point-biserial correlation.

c. Test for significance.

d. Write up the results.

6. A sample of 100 patients at the Colorado State Hospital were given an MMPI and an MCMI. The frequencies of patients who met criterion for paranoia subscales are indicated below:

	MMPI	
	Paranoid	Not Paranoid
Paranoid	62 (a)	17 (b)

MCMI

Not Paranoid	4 (c)	17 (d)

a. Set up H_o and H_a.

b. Perform an analysis using a phi correlation.

c. Test for significance.

d. Write up the results.

Intercorrelation Matrix Questions

Use the table on the page following to answer the questions below.

1. a. Which of the five subtests had the strongest correlation with Full Scale IQ?

   b. What was the r value for this correlation?

   c. Was it statistically significant?

   d. What was the level of the significance?

2. a. Which of the five subtests had the weakest correlation with Full Scale IQ?

   b. What was the r value for this correlation?

   c. Was it statistically significant?

   d. What was the level of the significance?

3. a. Which two of the five subtests had the strongest correlation with each other?

   b. What was the r value for this correlation?

   c. Was it statistically significant?

   d. What was the level of the significance?

4. a. If you were able to give only one subtest but you wanted the best predictor of Full Scale IQ, which subtest would you give?

   b. What was the r value for this correlation?

   c. Was it statistically significant?

   d. What was the level of the significance?

7. a. David Wechsler claimed that Vocabulary (VOC), Information (INF), and Digit Span (DSP) are verbal subtests and, thus, should be related to each other. He also claimed that Block Design (BD) and Digit Symbol (DSY) were performance subtests and, thus, should be related to one another. Does this correlation matrix support his contention?

   b. Discuss this intercorrelation matrix with regard to Wechsler's contention and your yes or no answer in five (a).

8. Discuss the 'problem' of Digit Span. Why could it be considered a problem? What would you recommend to Wechsler about it?

PEARSON CORRELATION COEFFICIENT

	VOC	BD	INF	DSP	DSY	FIQ
VOC	1.00 (N=41) p=.	.2498 (N=41) p=.058	.5319 (N=41) p=.000	.0182 (N=41) p=.455	.4055 (N=41) p=.004	.6298 (N=41) p=.000
BD	.2498 (N=41) p=.058	1.00 (N=42) p=.	.1960 (N=42) p=.107	.1747 (N=42) p=.134	.4462 (N=42) p=.002	.5962 (N=42) p=.000
INF	.5319 (N=41) p=.000	.1960 (N=42) p=.107	1.00 (N=42) p=.	−.0094 (N=42) p=.476	.3510 (N=42) p=.011	.4771 (N=42) p=.001
DSP	.0182 (N=41) p=.455	.1747 (N=42) p=.134	−.0094 (N=42) p=.476	1.00 (N=42) p=.	.0231 (N=42) p=.442	.1929 (N=42) p=.111
DSY	.4055 (N=41) p=.004	.4462 (N=42) p=.002	.3510 (N=42) p=.011	.0231 (N=42) p=.442	1.00 (N=42) p=.	.6353 (N=42) p=.000
FIQ	.6298 (N=41) p=.000	.5962 (N=42) p=.000	.4771 (N=42) p=.001	.1929 (N=42) p=.111	.6353 (N=42) p=.000	1.00 (N=42) p=.

7

The Statistical Analysis of the Archetypal Experiment: The *t* Test for Independent Groups

We previously learned that the correlation coefficient is probably the most often used inferential statistic. It has the limitation, however, that we can never imply a causal relationship between two correlated variables. We also mentioned earlier the very powerful archetypal experiment, where a large group of participants are randomly assigned to either an experimental group or a control group. The archetypal experiment does allow us to assume a causative relationship between the independent variable or treatment and the dependent variable or response variable that is measured on both groups. This experimental design and procedure is known and statistically analyzed by the **Independent Groups *t* test**.

The focus of the *t* test is to determine whether there is a significant difference between the experimental and control groups' means on the dependent variable beyond mere chance differences. In order to understand the focus of the *t* test, let us return to the signal detection theory analogy. If the independent variable really works, it is similar to creating a large signal because there will be a large difference between the experimental group's mean and the control group's mean. The magnitude of the difference between the two groups' means is considered to be the signal. If the independent variable really works, then the two groups' means will be very different; thus, the magnitude of the difference between the two groups will be large, and the signal will be large. This variation and difference in scores between the two groups because of the independent variable or treatment is called the **between-groups variance**.

Let us also imagine the situation where the independent variable does not work: In other words, nothing more than chance is having an influence upon the participants' scores. In this case, there should not be much of a difference between the experimental group's mean and the control group's mean beyond chance differences. These chance differences are similar to noise in signal detection theory. Notice also that each participant in the experimental group will vary from the group's mean. This variance is considered to be the within-group noise or **within-group variance**. In any experiment, the participants in a group will always vary from each other even though they were all in the same group, and even though they all received the same treatment. Statisticians refer to within-group variance as error or **within-group error**. In reality, it is rather bizarre to label any variation from participant to participant as an 'error' but it is considered to be an 'error' in name only.

In summary, the signal (the difference between the two groups' means) in an archetypal experiment must exceed the noise (within-group variation) in order for us to determine that the independent variable had a genuine effect upon the dependent variable. In terms of hypothesis testing, the true null and alternative hypotheses for the archetypal experiment would be:

$H_o: \mu_1 = \mu_2$

$H_a: \mu_1 \neq \mu_2$

These mean population values will actually be tested with sample means drawn from these population values. Thus, the actual null and alternative hypotheses tested will be:

$H_o: \bar{x}_1 = \bar{x}_2$

$H_a: \bar{x}_1 \neq \bar{x}_2$

ONE *t* TEST BUT TWO DESIGNS

There are actually two popular experimental designs that can be analyzed by the *t* test for independent groups. The first design is the archetypal experiment where the sample of participants is randomly selected from the population and then is randomly assigned to one of the two experimental groups, like an experimental group or a control group. Next, two levels of the independent variable are given or applied to the two groups, one level to one group, the other level to the other group. For example, the effects of mood state (good mood and bad mood) might be evaluated in helping behavior. Both groups' moods might be influenced by giving them false feedback on an ego-involving task. The Good Mood Group might be told they did very well on a test while the Bad Mood Group might be told they did very poorly. Next, each participant would be measured on how quickly he or she responds to help someone in need. Perhaps we might hire a graduate student (called a confederate) to pretend to be in need outside the experimental room where the participants had just been given their bogus feedback. If we found that the mean latency to respond was much shorter in the Good Mood Group compared to the Bad Mood Group, then we could preliminarily and cautiously conclude that the participants' moods did cause changes in helping behavior (depending, of course, on the results of the *t* test which tells us whether we can reject or retain H_o).

The second and common two-groups experimental design is sometimes called the *in situ* study. *In situ* is Latin and means 'in its original place.' In this design, the participants come as they are, pre-assigned by God or nature, to one of the two groups. For example, we might wish to determine whether children with dyslexia (reading difficulties) have IQs that are different from children without dyslexia. Thus, the two groups consist of dyslexic and non-dyslexic children. The presence or absence of dyslexia are the two levels of the independent variable. The dependent variable would be their IQ score. Notice that the participants are not randomly assigned to the two levels of the independent variable. However, the *in situ* experimental design does result in two independent groups and can be statistically analyzed with a *t* test exactly the same as the archetypal experiment. The limitation of the *in situ* experimental design is that we can be much less certain about causation, since a plethora of other variables might have accounted for the differences between the two groups. Despite this limitation, *in situ* studies are frequently the only way some psychological issues can be studied. For example, we cannot ethically randomly assign people to receive a head injury or not in order to determine whether head injuries affect memory loss. Thus, *in situ* experimental designs are popular in spite of the problem of not being able to determine causation.

ASSUMPTIONS OF THE INDEPENDENT *t* TEST

There are specific assumptions that must be met in order to use the *t* test appropriately. They are as follows:

Independent Groups: The participants must be different in each group, that is, no participant is allowed to serve in both groups. There is another form of the *t* test, the dependent groups *t* test, that does assume that the participants are the same in each group, and it will be discussed in Chapter 8.

Normality of the Dependent Variable: The *t* test and its critical values are based upon the assumption that the sample dependent variable values come from a population of values that is normally distributed. However, the *t* test's value is enhanced because it is still a fairly reliable measure even for non-normal but still mound shaped distributions. An interesting characteristic of the *t* test is that it is **robust**. The **robustness** of a statistical test means that its assumptions may be violated to some extent, yet the correct statistical decision will still be made, and that is, to correctly reject or retain the null hypothesis.

Homogeneity of Variance: While we may expect the means to be different between two groups in a *t* test, the assumption is made that the variances of the two groups about their respective means will be equal or approximately equal no matter whether the two groups' means are different or not. In reality this assumption is often a safe one. It is actually rare that the variances of two groups are radically unequal. Also, violations of the assumption matter more when the samples are small than when they are large. Thus, using larger sample sizes (like N > 15 or 20 in each group) helps to minimize unequal variances. Another way to reduce the effect of **heterogeneity of variance** is to use an equal number of participants in each group (also called **equal n**). The use of equal numbers of participants in each group has other beneficial statistical properties as well. Finally, the *t* test is robust against the violation of the assumption of homogeneity of variance and that means, despite violations of the assumption of homogeneity of variance, we are still likely to make the correct statistical decision.

THE FORMULA FOR THE INDEPENDENT t TEST

The formula for a t test between two different groups of scores is as follows:

$$t = \frac{\bar{x}_1 - \bar{x}_2}{\sqrt{\left[\frac{\Sigma x_1^2 - \frac{(\Sigma x_1)^2}{N_1} + \Sigma x_2^2 - \frac{(\Sigma x_2)^2}{N_2}}{N_1 + N_2 - 2}\right] \cdot \left[\frac{1}{N_1} + \frac{1}{N_2}\right]}}$$

where

$\bar{x}_1$ = the mean of the scores of the first group

$\bar{x}_2$ = the mean of the scores of the second group

Σx_1^2 = the sum of the squares of the first group

Σx_2^2 = the sum of the squares of the second group

$(\Sigma x_1)^2$ = the square of the sum of the scores of the first group

$(\Sigma x_2)^2$ = the square of the sum of the scores of the second group

N_1 = the total number of scores in the first group

N_2 = the total number of scores in the second group

YOU MUST REMEMBER THIS!: AN OVERVIEW OF HYPOTHESIS TESTING WITH THE t TEST

If the t value obtained by the formula exceeds the tabled critical value at the .05 significance level, then H_o will be rejected; it will then be concluded that one mean is significantly different from the other mean. If the t value obtained by the formula does not exceed the tabled critical value, then H_o will be retained, and it will be concluded that there is no significant difference between the two means other than just chance differences.

WHAT DOES THE t TEST DO?: COMPONENTS OF THE t TEST FORMULA

Let us examine the components of the t test formula. Mathematically, there are three major parts: the numerator, the left half of the denominator, and the right half of the denominator.

The numerator contains the difference between the two groups' means. What is the effect upon the final t value of the size of the numerator? It is easily seen that as the magnitude of the difference between the two means gets larger, the t value will get larger. In general, if the difference between the means is small, the t value will be small.

Now let us examine the left half of the denominator. After some of you recover from the shock, you will realize that this part of the formula looks vaguely familiar. It is the computational formula for the standard deviation. However, in this formula we have joined the standard deviations for the two groups together. This procedure results in what is called **pooled variance**. This means that the size of the variances for the two groups is approximately the same (or at least not radically different from each other) and, therefore they can be combined in a single formula. What is the effect of the size of the pooled variances upon the final t value? Since they are in the denominator, it means that, as they get larger, the t value gets smaller. If this is not intuitively obvious, imagine a pizza being divided among a family. If the family is large, the resulting pieces of pie will be small. If the family is smaller, then the pieces of pie will be larger. Thus, a large variance or standard deviation will reduce the size of the final t value. A small variance or standard deviation will enlarge the t value.

The final component of the t formula is the right half of the denominator. It contains the reciprocals of the sizes of the two groups. Why do we take the reciprocals? Taking the reciprocals reverses the effects of large and small values in the denominator. Normally, a large value in the denominator reduces the value of the final t. In this case, by taking the reciprocal, a large sample size increases the value of t, and a small sample size reduces the value of t.

In summary:

1. A large difference between the means creates a larger t value and, therefore, H_0 is more likely to be rejected.

2. Small variances or small standard deviations for the two groups creates a larger t value; therefore, H_0 is more likely to be rejected.

3. Large sample sizes create a larger t value; therefore, H_0 is more likely to be rejected.

WHAT IF VARIANCES ARE RADICALLY DIFFERENT FROM ONE ANOTHER?

One way to test statistically the assumption of homogeneity of variance is the F **Max test**, and it is described in advanced statistical texts. If the F Max test proves to be significant, it means that you must use a separate formula to obtain each variance or, in other words, you may not use the pooled variance formula. Another option is to transform each score by the same transformation formula and conduct the F Max test again. You may wish to consult an advanced statistics text should you fail to meet this assumption. However, we mentioned earlier that large sample sizes and equal numbers of participants in each group have the effect of minimizing violations of the assumption. In addition, the *t* test is robust.

A COMPUTATIONAL EXAMPLE

A sample of depressed patients was randomly assigned to either the experimental group or the control group: the experimental group was given Prozac (a popular anti-depressant) and the other group was given a placebo. Neither of the groups' members knew whether they were taking Prozac or a placebo. After one month of drug treatment, each patient completed the Coolidge Axis II Inventory Depression scale on which scores can range from 24 to 96 and scores above 60 are indicative of a clinical depression.

Prozac group	Placebo group
53	64
44	68
61	84
50	65
40	50
39	60
58	81
42	76

As far as the formula is concerned, either group can be labeled Group 1. Although it does not matter, if you wish to have a positive *t* value, then label the group with the highest mean as Group 1. In this example, the placebo group appears to have the highest mean, and it will be labeled Group 1, while the Prozac group will be Group 2.

Steps in the *t* test formula:

Step 1. Calculate $\bar{x}_1$

$$\bar{x}_1 = \frac{64 + 68 + 84 + 65 + 50 + 60 + 81 + 76}{8}$$

$$\bar{x}_1 = \frac{548}{8}$$

$$\bar{x}_1 = 69.500$$

Step 2. Calculate $\bar{x}_2$

$$\bar{x}_2 = \frac{53 + 44 + 61 + 50 + 40 + 39 + 58 + 42}{8}$$

$$\bar{x}_2 = \frac{387}{8}$$

$$\bar{x}_2 = 48.375$$

Step 3. Calculate Σx_1^2

$$\Sigma x_1^2 = 64^2 + 68^2 + 84^2 + 65^2 + 50^2 + 60^2 + 81^2 + 76^2$$

$$\Sigma x_1^2 = 4096 + 4624 + 7056 + 4225 + 2500 + 3600 + 6561 + 5776$$

$$\Sigma x_1^2 = 38,441$$

Step 4. Calculate Σx_2^2

$$\Sigma x_2^2 = 53^2 + 44^2 + 61^2 + 50^2 + 40^2 + 39^2 + 58^2 + 42^2$$

$$\Sigma x_2^2 = 2809 + 1936 + 3721 + 2500 + 1600 + 1521 + 3364 + 1764$$

$$\Sigma x_2^2 = 19,215$$

Step 5. Calculate $(\Sigma x_1)^2$

$$(\Sigma x_1)^2 = (64 + 73 + 84 + 65 + 50 + 60 + 81 + 76)^2$$
$$(\Sigma x_1)^2 = (548)^2$$
$$(\Sigma x_1)^2 = 300,304$$

Step 6. Calculate $(\Sigma x_2)^2$

$$(\Sigma x_2)^2 = (53 + 44 + 61 + 50 + 40 + 39 + 58 + 42)^2$$
$$(\Sigma x_2)^2 = (387)^2$$
$$(\Sigma x_2)^2 = 149,769$$

Step 7. Obtain the number of scores in each group.

$N_1 = 8$ (the number of scores in Group 1)

$N_2 = 8$ (the number of scores in Group 2)

Step 8. Enter the values obtained in steps 1–7 into the formula for the *t* test.

$$t = \frac{\bar{x}_1 - \bar{x}_2}{\sqrt{\left[\dfrac{\Sigma x_1^2 - \dfrac{(\Sigma x_1)^2}{N_1} + \Sigma x_2^2 - \dfrac{(\Sigma x_2)^2}{N_2}}{N_1 + N_2 - 2}\right] \cdot \left[\dfrac{1}{N_1} + \dfrac{1}{N_2}\right]}}$$

$$t = \frac{68.500 - 48.375}{\sqrt{\left[\dfrac{38{,}438 - \dfrac{(548)^2}{8} + 19{,}215 - \dfrac{(387)^2}{8}}{8 + 8 - 2}\right] \cdot \left[\dfrac{1}{8} + \dfrac{1}{8}\right]}}$$

$$t = \frac{20.1250}{\sqrt{\left[\dfrac{38{,}438 - 37{,}538 + 19{,}215 - 18{,}721.125}{14}\right] \cdot \left[\dfrac{1}{4}\right]}}$$

$$t = \frac{20.1250}{\sqrt{\left[\dfrac{1{,}393.875}{14}\right] \cdot [0.25]}}$$

$$t = \frac{20.1250}{\sqrt{24.890625}}$$

$$t = \frac{20.1250}{4.9890505}$$

$$t = 4.034$$

TESTING THE NULL HYPOTHESIS

In order to test the null hypothesis, the derived t value will be compared to **critical values** of t on the **t distribution**. Refer to the table of critical t test values in Appendix B.

Steps in Determining Significance

Step 1. In order to use the table of t values, the degrees of freedom (df) are needed. The df formula is $N_1 + N_2 - 2$ or in the previous example, $8 + 8 - 2 = 14$. The concept of df is a complex one, but one simple aspect is that the df for a t test is roughly related to the total number of participants, but is always less.

Step 2. Determine whether you will conduct a one-tailed or two-tailed test of significance. If you have chosen a non-directional alternative hypothesis (as has been suggested throughout the book), then you have already decided on a two-tailed test of significance. In our example, we had a non-directional alternative hypothesis; therefore, we have chosen a two-tailed test of significance.

Step 3. Determine the level of significance at which you will conduct the test of the null hypothesis. Statisticians are in nearly complete agreement that the starting level of significance should be $p = .05$ (this starting level is also called **the conventional level of significance**). Therefore, the null hypothesis will be tested against the conventional level of $p = .05$.

Step 4. Determine the critical values of t at $p = .05$ with $df = N_1 + N_2 - 2$. In our example, for a two-tailed test with $p = .05$ and $df = 14$, Appendix B reveals that the critical values of $t = \pm 2.145$.

Step 5. Finally, compare the formula derived value of t to the critical values of t. If the t value does exceed this tabled value, then the null hypothesis is rejected, and it is concluded that the means are significantly different from each other. If the t value does not exceed the tabled value at the .05 level of significance, then the null hypothesis is retained, and it is concluded that there is no significant difference between the means (this does not imply that the means are equal, only that they are not significantly different from each other).

Step 6. Report the findings. The discipline of psychology, as well as over 200 journals, have adopted the American Psychological Association format (APA, 1994) for the reporting of statistical tests. Their recommendations are to report the test statistic (r or t), its derived value, the degrees of freedom, and the significance level (p level).

In our example, the null and alternative hypotheses are:

$H_o: \bar{x}_1 = \bar{x}_2$

$H_a: \bar{x}_1 \neq \bar{x}_2$

where

$\bar{x}_1$ = control or placebo group

$\bar{x}_2$ = experimental or Prozac group

The derived $t = 4.034$ exceeds the critical value of $t = \pm 2.145$ at $p = .05$, with $df = 14$. Therefore, H_o is rejected, and it is concluded that the mean CATI Depression score for the Prozac group (48.4) was significantly lower than the mean of the placebo group (68.5), $t(14) = 4.034$, $p < .01$. In terms of the research question, it appears that Prozac relieves depression.

When H_o has been Rejected: When H_o has been rejected, the standard statistical procedure is to report the lowest p level (also known as the alpha level). In the present case, H_o was rejected at the .05 level but the derived t value also exceeded the critical value of t at $p = .01$ but it did not exceed the critical value at $p = .001$. Therefore, the t will be reported as statistically significant at $p < .01$.

Note that this significance procedure is counter-intuitive. This means that the derived t value must **exceed** the tabled critical value of t in order to report it significant at **less than** .05. Remember, however, what this testing process represents. If there is no effect of the independent variable between the two groups (the treatment doesn't work), then theoretically the derived t value should be zero or very close to zero. If the treatment does work, then the two groups' means should be far apart, creating a large t value. Thus, if the independent variable does have an effect on the two groups' means, the null hypothesis will be rejected because the derived t value is likely to be very large.

THE POWER OF A STATISTICAL TEST

The **power** of a statistical test refers to its ability to detect a false null hypothesis or its ability to detect a real difference between two groups' means. In general, the independent t test is a powerful test because it does have a solid ability to detect true treatment differences between two means. However, the power of a test can be experimentally manipulated, and the power of a statistical test can be underused or abused. The power of a statistical test can be underused if we fail to use enough participants. In this case, although there may be a real difference between two groups' means (the treatment really works), we may fail to reject the null hypothesis (committing the Type II Error) if we do not have enough participants in each group. Remember an experiment is psychometrically large if the total N exceeds 30. Thus, to have sufficient power it is recommended that each group have at least 15 participants. Another way of experimentally increasing the power of a test would be to try to make the participants as homogeneous as possible (like restricting the ages of the participants). While this might have the effect of limiting the generality of the results, it may also have the effect of reducing the within-groups variance or noise in the experiment, and this will increase the power of the statistical test.

It is also common in the social sciences to witness an abuse of power with statistical tests. In this case, the abuse comes from having too many participants in the study. While it may appear that 'the larger, the better' or 'the more, the merrier' may be true in surveys and sampling, in inferential statistics like correlation and the independent t test, too many participants may result in an abuse of statistical power. Using more participants than needed does increase the chance of detecting a real difference between two means. Thus, using a lot of participants does reduce the chance of committing the Type II Error (retaining H_o when it's false). However, there is a cost for this increased detection ability and that is we are much more likely to commit the Type I Error when we use too may participants. This occurs because the increased sensitivity to real differences also makes us vulnerable to claiming chance differences as real differences. The solution to the power dilemma is to conduct a **power analysis**. A power analysis determines how many participants would be appropriate for a study. Power analysis tends to be an advanced topic in statistics, so consult an advanced statistical text.

EFFECT SIZE

Remember when we ran into the curious artifact in correlation where a correlation might be weak (even as low as $r = .06$) and still be statistically significant or a correlation could be strong (as high as $r = .56$), yet not be statistically significant? This curiosity often occurs as an interaction between the power of a statistical test and the effect size. **Effect size** refers to the effect of the influence of independent variable upon the dependent variable. Another way to consider effect size is how well the treatment works. If the treatment really works (as detected by a large difference between the two groups' means), then there is said to be a large effect size. If the difference between the two groups' means is small, then there is said to be a small effect size. Notice the same curious dilemma as we witnessed with correlation: we can have a strong effect size but fail to reject H_0. This may occur because we failed to have sufficient statistical power (like too few participants in each group). We can also have a very weak effect size and, yet, we may still reject H_0. This may occur in cases where we have abused statistical power by using too many participants. Once again the solution to the effect size dilemma is to conduct a power analysis.

THE CORRELATION COEFFICIENT OF EFFECT SIZE

In order to determine the effect size in the independent t test, a correlation coefficient of effect size can be derived. The formula is:

$$r = \sqrt{\frac{t^2}{t^2 + df}}$$

The correlation coefficient of effect size will always be positive and range from 0 to 1.00. Use the following scale to interpret the magnitude of the effect size:

Effect size	r value
small	.100
medium	.243
large	.371

Notice that these r values are considerably smaller than we had for the same labels when we used the Pearson Product-Moment Correlation Coefficient r. Thus, keep in mind that these are two separate statistical procedures, although both use the coefficient r.

In the previous example, we obtained a t value of 4.034. If we enter this value into the correlation coefficient of effect size formula, we obtain the following:

$$r = \sqrt{\frac{t^2}{t^2 + df}}$$

$$r = \sqrt{\frac{4.034^2}{4.034^2 + 14}}$$

$$r = .733$$

Thus, we can see that a very large effect size was observed in the Prozac experiment or, in other words, there was a major effect of Prozac in reducing depression in these clinically depressed patients.

HISTORY TRIVIA

William Gossett (1876–1937) is attributed with the discovery of the t distribution. He worked part-time for the Guinness Brewery in Ireland while pursuing his graduate studies. Later, he was employed by the brewery to do research, and he discovered that using small samples in sampling procedures led to errors when trying to estimate population values. This led him to the discovery of the family of t distributions based on varying small sample sizes. In 1908, he published an article about his findings, but Guinness Brewery had strict regulations about publications by its employees, so Gossett chose a pen name, 'Student.' It is not exactly known why he chose to label the new distributions 't,' although it may have had something to do with 'tea' breaks in the afternoon.

By 1920, Student had published tables of the **t distribution** and the **t statistic**. In 1925, Ronald A. Fisher (1890–1962) published a book called *Statistical Methods for Research Workers*. One major contribution of this work was the idea of using Student's t statistic to test a hypothesis about a single mean and to test the difference between two means. Thus, the **t test** was actually created by Fisher based on Gossett's family of t distributions.

KEY SYMBOLS AND TERMS

---

t test – The t test for independent groups determines whether there are significant differences between two independent groups' means on the same dependent variable.

Archetypal Experiment – The archetypal experiment is the most powerful experimental design because it allows for the implication of causation. The participants are randomly chosen from a population and randomly assigned to one of two groups, usually an experimental group and a control group that receives a placebo.

In Situ Design – *In situ* means 'in its original place,' and it is an often used *t* test experimental design where the participants are not randomly assigned to the two groups but have been pre-assigned by nature or God, like dyslexic children and non dyslexic children, or brain-injured patients and controls, etc.

Homogeneity of Variance – An assumption in the independent groups *t* test is that the two groups have equal or approximately equal variances.

Pooled Variance – The denominator of the independent *t* test contains the pooled variance estimate. If the assumption of homogeneity of variance has been met, then the two groups' variances may be pooled together to make a single estimate. If the assumption of homogeneity has not been met then separate estimates of variance must be conducted for the two groups.

F Max Test – A statistical test used to assess the homogeneity of variance assumption in the independent *t* test. A significant *F* Max test indicates that the two groups' variances are significantly different.

Robustness – A characteristic of statistical tests which indicates the ability to detect a real difference between the two groups' mean or the ability to detect a false null hypothesis despite violations of statistical assumptions.

Degrees of Freedom – A complicated statistical term which in some statistical tests is roughly correlated with the total number of participants or observations but always slightly less. The *df* is actually based in the estimation of the standard deviation and indicates the number of numbers that are free to vary in estimation theory.

Power – The power of a statistical test refers to the probability that the null hypothesis will be rejected correctly.

Effect Size – The effect size in the independent *t* test refers to how strongly the independent variable affected the dependent variable.

1. & 2. A researcher wishes to determine whether Vitamin C prevents colds and whether the number of days sick is reduced by Vitamin C. Thirty subjects were randomly assigned to either the Vitamin C group or the placebo group. Are there mean differences between the groups on their number of colds in a 3 year period? Are there mean differences in the number of days sick?

Number of colds in 3 years		Number of days sick	
Vit-C	Placebo	Vit-C	Placebo
3	8	9	24
8	8	16	32
6	7	24	21
7	10	18	40
4	11	17	37
9	4	25	20
2	3	3	15
5	7	17	73
7	6	20	33
11	8	6	37
10	4	28	19
8	10	22	62
7	6	20	33
6	8	21	39
7	5	18	35

3. What is the purpose of the t test?

4. Explain the assumption of homogeneity of variance.

5. What can be done to reduce the effects of heterogeneity of variance?

6. Describe the three factors that can make a t value larger.

7. Why is it counter-intuitive to reject H_o if the derived t value exceeds the tabled critical value of t?

8. Explain the concept of the power of a statistical test. How can it be misused?

9. If a t value of 3.587 is obtained with 10 degrees of freedom, what will be the reported p level of significance?

10. What happens if the *F* Max test is significant?

11. A researcher assessed the effects of a new drug for the control of anxiety. Eighteen anxious patients were randomly assigned to one of two groups, either the drug group or the placebo group. One half hour after the administration of the drug or placebo, their anxiety was measured and reported as T scores (mean = 50, SD = 10). Does the new drug seem to reduce anxiety?

Drug	Placebo
52	65
53	59
58	68
50	53
53	59
58	67
55	61
66	70
53	62

12. A researcher wishes to determine whether women who have been receiving estrogen replacement therapy die at an older or an earlier age than their sisters who have not received estrogen. Estrogen groups' ages at death: 81, 72, 84, 78, 85, 82, 79, 83. No estrogen groups' ages at death: 71, 79, 84, 66, 69, 70, 77, 83. Is there a difference in the two groups' ages?

8 Variations on the Archetypal Experiment: The *t* Test for Dependent Groups

The *t* test for dependent groups (or **dependent *t* test**) is used to analyze the difference between two groups' means in experimental designs where the participants in both groups are related to each other in some way. The two most common designs are where the same participant serves in both groups or where pairs of matched subjects are split between the two groups. The dependent *t* test has an interesting number of names, despite a limited number of designs. For example, this test has also been called the **repeated measures *t* test, dependent-samples *t* test, within-subjects *t* test, paired *t* test, matched *t* test, paired difference *t* test,** and **correlated *t* test**. Although the independent *t* test is far more common than the dependent t, the dependent *t* test can, in many circumstances, be a more powerful test than the independent *t* test. Remember that the power of a statistical test is its ability to detect a genuine difference between two means. Before we examine the reasons for the increased power of the dependent *t* test, let us review the three essential experimental designs of the dependent *t* test in order of their prevalence.

DESIGN 1

The same participant is used in both groups but the presentation of the independent variable is fixed by God or nature.

Example of Design 1. This design is the most common of the three dependent *t* test designs. It occurs when any participants are measured before and after some treatment, and their 'before' scores are compared to their 'after' scores. It is also referred to as the 'pre- and post-treatment' design. For example, we might compare subject level of distress before psychotherapy and then measure their distress after psychotherapy. If the distress scores go down after psychotherapy, we may tentatively and cautiously conclude that psychotherapy may have been the cause of the decline in distress scores. However, this is a very tenuous assumption. Since we were not able to randomly present the independent variable, a slew of other factors may have actually been responsible like the simple passage of time. Another possibility is the **regression effect** where extreme scores when measured again tend to 'regress' towards the mean. Because there may have been factors other than the treatment effects causing the difference in before and after experimental t designs, we must be very cautious about any implications of causation.

DESIGN 2

Different participants are used in both groups but the participants are paired or matched on some critical variables (in order to make them as alike as possible). The matching critical variables must be known to be related to the dependent variable.

Example of Design 2. This design is not as popular as the previous design. For example, imagine a study where the researcher is interested in long-term memory. The researcher knows that long-term memory varies widely among people, so in order to control for this variability, the researcher employs the matched t test design because the variability across the same participant over time will be less than the variability of long-term memory between two different participants. The researcher wishes to know whether visual imagery leads to better long-term memory or whether simple rehearsal of the material leads to better long-term memory. If the researcher decides to match the participants, it had better be on variables critical to long-term memory. One variable critical to better long-term memory is age and even more critical is IQ. Therefore, the researcher should match the participants on age and IQ as carefully as possible in order to control for these effects upon the dependent variable. Thus, for each matched pair, one will receive the imagery condition and one will be in the rehearsal condition. The dependent t test will compare the means of the two types of memory enhancement for the amounts of memory retained after a set period of time.

One major drawback to this design is that the researcher may not often know nor be able to control for all of the critical variables related to the dependent variable. Also remember that for all three of these designs, there must be an equal number of scores in each group since the scores represent either the same subject in each group or matched pairs of subjects.

DESIGN 3

The same participant is used in both groups. The independent variable is balanced in presentation across all of the participants (for half of the participants, they receive Level 1 of the independent variable first, and for the other half of the participants, they receive Level 2 of the independent variable first).

Example of Design 3. This design is far less popular than either of the previous two designs. In this design one concern is the order of presentation of the independent variable. The **order of presentation** of the two levels of an independent variable may affect the results of an experiment since the participants may suffer from fatigue or boredom after a period of time. Thus, if we can balance the presentation of the two levels of an independent variable in an experiment, we can experimentally control for the order of presentation effect. With the Wechsler Adult Intelligence Scale, it was noticed that most people scored lower on the five Performance subtests than on the six verbal subtests. In part this finding may have been due to order of presentation since all six Verbal subtests were administered first (often taking an hour) before the five Performance subtests. When the order of presentation was balanced across the Verbal and Performance subtests it was found that the magnitude of the discrepancy decreased but the difference did persist.

ASSUMPTIONS OF THE DEPENDENT *t* TEST

The assumptions of the dependent *t* test are similar to the independent *t* test. The dependent variable is assumed to come from a population of scores that is normally distributed. As noted earlier, there must also be an equal number of scores in each group, although this is an experimental design requirement as well as a statistical requirement. It is also assumed that the scores within a group are independent of one another. As an example of a violation of the latter assumption, imagine if as one tested participants, one's laboratory became hotter and smellier. The participants tested later might have scores that were affected by the previous participants, thus, violating the assumption of independence of scores within a group.

WHY THE DEPENDENT *t* TEST MAY BE MORE POWERFUL THAN THE INDEPENDENT *t* TEST

As we mentioned earlier, the variability of the same participant over time is less than the variability between two unrelated participants. For example, imagine pairing your IQ this year and next year to two unrelated people's IQs. There, of course, will be greater variability between two unrelated people's IQ scores. Also, imagine pairing the IQs of two identical twins (matched pairs) compared to the variability of two unrelated people's IQs. Once again there will be less variability between the matched pairs of scores than two unrelated scores. Thus, the dependent *t* test design is more powerful because it can reduce the noise in an experiment (or better known as the within-subject variance or within-subject error).

HOW TO INCREASE THE POWER OF A *t* TEST

There are at least three ways to increase the power of an independent *t* test or dependent *t* test. One way is to increase the difference between the two means, perhaps by creating a stronger treatment effect. In the example of the memory enhancement experiment, the treatment may be strengthened by training the participants in the visual imagery and rehearsal conditions for a longer period of time, like instead of just one session of training, train the participants for five sessions. In the pre- and post-treatment example with psychotherapy, the length of psychotherapy might be increased from one month of stress reduction to three months. Another way to increase the power of a t test is to increase the number of participants in a study. Remember that a statistically large sample begins at $N = 30$. Yet also remember that in inferential statistics we can abuse power by using far too many participants. A power analysis can determine the appropriate number of participants in a study. A third way of increasing power is to decrease the within-subject variance in each group. The dependent t test automatically decreases the within-subject variance because either the participants are used twice (under both conditions of the independent variable) or variance is reduced by matching the participants on critical variables. Additionally, within-subject variance may be reduced by trying to homogenize the participants, that is, by using similar groups like only ages 18–21, middle-class, etc. Homogenization does have the negative effect of restricting generalizations from the results to other populations but the power of the experiment to detect genuine differences in the data is enhanced.

DRAWBACKS OF THE DEPENDENT t TEST DESIGNS

A word of caution is in order. The dependent t test does have some drawbacks. One issue deals with the concept of degrees of freedom. Remember that the formula for degrees of freedom in the independent t test is $N_1 + N_2 - 2$. In a dependent t test, the dependence between the pairs of scores means that knowing a score under one condition fixes (to some extent) the score in the other condition. For example, if a researcher has used twins in an IQ study, knowing one twin's IQ means that the other twin's IQ in the second condition is not really free to vary but will be highly correlated to his or her matching twin's IQ score. Thus, the formula for the degrees of freedom in the dependent t test is half of the degrees of freedom in the independent t test, and it is $N - 1$ where N is the *number of pairs* of scores. This reduction in the number of degrees of freedom actually reduces the statistical power of the dependent t test since the reduction produces a higher critical value of t in the family of t distributions in Appendix B. The higher critical value of t makes it more difficult to reject the null hypothesis.

ONE-TAILED OR TWO-TAILED TESTS OF SIGNIFICANCE

The way in which the alternative hypothesis is stated determines whether a test of significance is one-tailed or two-tailed. It is one-tailed if the alternative hypothesis has been stated directionally (for example, the pre-test mean will be less than the post-test mean). It is two-tailed if the alternative hypothesis is stated non-directionally (for example, there will be a difference between the pre-test and post-test means). Each test of significance has its advantages and disadvantages. The one-tailed test of significance is more sensitive to genuine differences between the two means, and it is also more sensitive to smaller real differences between the two means. However, the cost of this sensitivity is that it is also more sensitive to chance differences and to interpreting chance differences as real differences (the Type I Error).

The two-tailed test of significance protects against the Type I Error better than the one-tailed test of significance. It is less likely to have us interpret chance differences between two means as being real differences. However, this protection comes at a cost and that is it will be less sensitive to smaller but genuine differences between two means. Many statisticians feel that the non-directional alternative hypothesis provides sufficient sensitivity to real differences and provides better protection against interpreting chance differences as real differences than the one-tailed test. Also, I have two other thoughts about directional alternative hypotheses. First, it is not uncommon in scientific research to have a finding the exact opposite of what was predicted. The one-tailed test of significance puts the entire rejection region (.05) in a single tail of the t distribution. If the results are the opposite of what has been predicted, we are forced to retain the null hypothesis since we can only reject the null hypothesis if the direction of the finding is as we predicted. If it is the opposite of what we predicted, we have no alternative but to retain the null hypothesis in the one-tailed test. My second thought is that if one is so certain of the outcome of an experiment, why is one doing the experiment in the first place? In fact, two-tailed tests are so prevalent in research that it has been often wondered whether in experiments that report one-tailed tests of significance, the experimenters failed to attain significance with a two-tailed test and then the experimenters switched to a one-tailed test in order to attain significance.

HYPOTHESIS TESTING AND THE DEPENDENT t TEST: DESIGN 1

The null and non-directional alternative hypotheses for Design 1 of the dependent t test is as follows:

H_o: $\mu_1 = \mu_2$ or $\mu_1 - \mu_2 = 0$

H_a: $\mu_1 \neq \mu_2$ or $\mu_1 - \mu_2 \neq 0$

The formula for the dependent t test is:

$$t = \frac{\bar{x}_1 - \bar{x}_2}{\sqrt{\frac{\Sigma D^2 - \frac{(\Sigma D)^2}{N}}{N(N-1)}}}$$

where

$\bar{x}_1$ = the mean of the pre-test scores

$\bar{x}_2$ = the mean of the post-test scores

ΣD^2 = the sum of the squares of the differences between the pre-test scores and the post-test scores

$(\Sigma D)^2$ = the square of the sum of the differences between the pre-test scores and the post-test scores

N = the number of *pairs* of scores

Design 1 (Same Participants or Repeated Measures): A Computational Example

Patients in high distress entered into psychotherapy and were assessed for their stress levels before and after three psychotherapy sessions. The purpose of the study was to determine whether a new type of brief psychotherapy was successful after just three sessions. The patients' distress was measured by the Distress Scale of the Coolidge Axis II Inventory (and the patients' scores are reported as T scores).

	Distress scores (as T score values)	
	Before therapy	After therapy
Patient 1	68	63
Patient 2	58	60
Patient 3	74	65
Patient 4	55	62
Patient 5	81	54
Patient 6	59	73
Patient 7	47	45
Patient 8	75	73

Step 1. Subtract the pairs of scores from each other in the following manner.

	Before		After	Difference scores (D scores)
Patient 1	68	–	63	= 5
Patient 2	58	–	60	= –2
Patient 3	74	–	65	= 9
Patient 4	55	–	62	= –7
Patient 5	81	–	54	= 27
Patient 6	59	–	73	= –14
Patient 7	47	–	45	= 2
Patient 8	75	–	73	= 2

Step 2. Calculate the mean of the pre-test scores ($\bar{x}_1$).

$$\bar{x}_1 = \frac{68 + 58 + 74 + 55 + 81 + 59 + 47 + 75}{8}$$

$$\bar{x}_1 = \frac{517}{8}$$

$$\bar{x}_1 = 64.625$$

Step 3. Calculate the mean of the post-test scores ($\bar{x}_2$).

$$\bar{x} = \frac{63 + 60 + 65 + 62 + 54 + 73 + 45 + 73}{8}$$

$$\bar{x}_2 = \frac{495}{8}$$

$$\bar{x}_2 = 61.875$$

Step 4. Calculate the sum of the squares of the differences between the pre-test scores and the post-test scores ($\Sigma D)^2$.

$$\Sigma D^2 = 5^2 + (-2)^2 + 9^2 + (-7)^2 + 27^2 + (-14)^2 + 2^2 + 2^2$$

$$\Sigma D^2 = 25 + 4 + 81 + 49 + 729 + 196 + 4 + 4$$

$$\Sigma D_2 = 1{,}092$$

Step 5. Obtain the square of the sum of the differences between the pre-test scores and the post-test scores.

$$(\Sigma D)^2 = (5 - 2 + 9 - 7 + 27 - 14 + 2 + 2)^2$$

$$(\Sigma D)^2 = (22)^2$$

$$(\Sigma D)_2 = 484$$

Step 6. Determine the number of pairs of scores. In the formula, N refers to the number of pairs of scores.

$$N = 8$$

Step 7. Enter the values obtained from steps 1–6 into the formula for the dependent test.

$$t = \frac{\bar{x}_1 - \bar{x}_2}{\sqrt{\dfrac{\Sigma D^2 - \dfrac{(\Sigma D)^2}{N}}{N(N-1)}}}$$

$$t = \cfrac{64.625 - 61.875}{\sqrt{\cfrac{10{,}092 - \cfrac{22^2}{8}}{8(8-1)}}}$$

$$t = \frac{2.75}{18.4196}$$

$$t = 0.641$$

Step 8. Compare the derived t value to the critical values of t in Appendix B. The formula for the df in a dependent t test is $df = N - 1$, where $N =$ the number of pairs of scores. In the present example, there were 8 pairs so $df = 8 - 1$, or $df = 7$.

Step 9. Determine whether the H_o should be retained or rejected by comparing the derived t value to the tabled critical values of t at $p = .05$ with $df = 7$. If the derived t value exceeds the tabled critical value of t (or if it is less than the negative t critical value), then H_o is rejected. If not, then H_o is retained.

The null and alternative hypotheses are:

$H_o: \mu_{\text{pre-test}} = \mu_{\text{post-test}}$

$H_a: \mu_{\text{pre-test}} \neq \mu_{\text{post-test}}$

Step 10. Write-up your findings and make a conclusion.

Thus, the derived $t = 0.641$ does not exceed the tabled critical value of $t = \pm$ 2.365 at $p = .05$ with $df = 7$. Therefore, H_o is retained, and it is concluded that the mean Distress scale T score after therapy (61.9) was not significantly lower than the T scores before psychotherapy (64.6), $t(7) = 0.641$, $p > .05$. In terms of the research question, it appears that this type of psychotherapy did not appear to be effective in this sample of distressed patients.

Determination of Effect Size

As noted earlier, the significance of a statistical test presents only part of the picture. Effect size estimation is useful for determining the strength of the independent variable, regardless of whether H_o was retained or rejected. In the present example, the effect size determination would be:

$$r = \sqrt{\frac{t^2}{t^2 + df}}$$

$$r = \sqrt{\frac{0.641^2}{0.641^2 + 7}}$$

$$r = .31$$

By comparing this value to the table of effect sizes on page 155, we see that it translates into a medium size effect. In essence, this determination of effect size indicates that despite a lack of statistical significance, there appears to be a medium strength effect of psychotherapy in reducing the distress of this sample of patients.

Design 2 (Matched Pairs): A Computational Example

A researcher wishes to determine which of two methods of memory training works best with the elderly. Twins over the age of 65 were selected for the study, half were trained to memorize a list of words using visual imagery and the other half spent the same amount of time simply memorizing the words. The number of correct words (maximum 20 correct) recalled was measured in both groups 24 hours later. The results are as follows:

Correct number of words			
Visual imagery		Simple rehearsal	
Twin 1a	15	Twin 1b	12
Twin 2a	10	Twin 2b	11
Twin 3a	12	Twin 3b	9
Twin 4a	9	Twin 4b	7
Twin 5a	16	Twin 5b	13
Twin 6a	18	Twin 6b	15
Twin 7a	11	Twin 7b	12
Twin 8a	14	Twin 8b	12

Step 1. Subtract the pairs of scores from each other in the following manner.

Visual imagery		Simple rehearsal		Difference scores (D scores)
Twin 1a	15	Twin 1b	12	15 – 12 = 3
Twin 2a	10	Twin 2b	11	10 – 11 = –1
Twin 3a	12	Twin 3b	9	12 – 9 = 3
Twin 4a	9	Twin 4b	7	9 – 7 = 2
Twin 5a	16	Twin 5b	13	16 – 13 = 3
Twin 6a	18	Twin 6b	15	18 – 15 = 3
Twin 7a	11	Twin 7b	12	11 – 12 = –1
Twin 8a	14	Twin 8b	12	14 – 12 = 2

Step 2. Calculate the mean of the Visual Imagery condition ($\bar{x}_1$).

$$\bar{x}_1 = \frac{15 + 10 + 12 + 9 + 16 + 18 + 11 + 14}{8}$$

$$\bar{x}_1 = \frac{105}{8}$$

$$\bar{x}_1 = 13.125$$

Step 3. Calculate the mean of the Simple Rehearsal condition ($\bar{x}_2$).

$$\bar{x}_2 = \frac{12 + 11 + 9 + 7 + 13 + 15 + 12 + 12}{8}$$

$$\bar{x}_2 = \frac{91}{8}$$

$$\bar{x}_2 = 11.375$$

Step 4. Calculate the sum of the squares of the differences between the pre-test scores and the post-test scores $(\Sigma D)^2$.

$$\Sigma D^2 = 3^2 + (-1)^2 + 3^2 + 2^2 + 3^2 + 3^2 + (-1)^2 + 2^2$$

$$\Sigma D^2 = 9 + 1 + 9 + 4 + 9 + 9 + 1 + 2$$

$$\Sigma D^2 = 44$$

Step 5. Obtain the square of the sum of the differences between the pre-test scores and the post-test scores

$$(\Sigma D)^2 = (3 - 1 + 3 + 2 + 3 + 3 - 1 + 2)^2$$

$$(\Sigma D)^2 = (14)^2$$

$$(\Sigma D)^2 = 196$$

Step 6. Determine the number of pairs of scores. In the formula, N refers to the number of pairs of scores.

$$N = 8$$

Step 7. Enter the values obtained from steps 1–6 into the formula for the dependent t test.

$$t = \frac{\bar{x}_1 - \bar{x}_2}{\sqrt{\dfrac{\Sigma D^2 - \dfrac{(\Sigma D)^2}{N}}{N(N-1)}}}$$

$$t = \frac{13.125 - 11.375}{\sqrt{\dfrac{44 - \dfrac{14^2}{8}}{8(8-1)}}}$$

$$t = \frac{1.75}{0.5901}$$

$$t = 2.966$$

Step 8. Compare the derived t value to the critical values of t in Appendix B. The formula for the df in a dependent t test is $df = N - 1$, where $N =$ the numbers of scores. In the present example, there were 8 pairs so $df = 8 - 1$, or $df = 7$.

Step 9. Determine whether the H_o should be retained or rejected by comparing the derived t value to the tabled critical values of t at $p = .05$ with $df = 7$. If the derived t value exceeds the tabled critical value of t (or if it is less than the negative t critical value), then H_o is rejected. If not, then H_o is retained.
The null and alternative hypotheses are:

$H_o: \mu_{visual} = \mu_{rehearsal}$

$H_a: \mu_{visual} \neq \mu_{rehearsal}$

Step 10. Write-up your findings and make a conclusion.
Thus, the derived $t = 2.966$ exceeds the tabled critical value of $t = \pm 2.365$ at $p = .05$ with $df = 7$. Therefore, H_o is rejected, and it is concluded that the mean number of words correct under the Visual Imagery condition (13.1) was significantly greater than the mean number of words correct in the Simple Rehearsal condition (11.4), $t(7) = 2.966, p < .05$. In terms of the research question, it appears that visual imagery appears to enhance memory recall better than simple rehearsal in this sample of twins.

Determination of Effect Size
The effect size determination would be:

$$r = \sqrt{\frac{t^2}{t^2 + df}}$$

$$r = \sqrt{\frac{2.966^2}{2.966^2 + 7}}$$

$$r = .83$$

By comparing this value to the table of effect sizes on page 151, we see that it translates into a very large size effect. Thus, it appears not only is visual imagery statistically significantly superior to simple rehearsal but the strength of this effect is considerably large.

Design 3 (Same Participants and Balanced Presentation): A Computational Example

A researcher wished to determine whether consistently higher verbal intelligence (VIQ) scores than performance intelligence (PIQ) scores are due to order of presentation effects on the Wechsler Adult Intelligence Scale (WAIS) since the six subtests of the verbal portion are all administered before the five subtests on the performance section. The researcher counterbalances the presentation of the verbal and performance subtests, assigning half of the participants to take the verbal portion first and half of the participants to take the performance subtests first. The data are tabulated as follows:

Participant 1	VIQ	112	PIQ	109
Participant 2	VIQ	117	PIQ	110
Participant 3	VIQ	99	PIQ	105
Participant 4	VIQ	128	PIQ	114
Participant 5	PIQ	103	VIQ	120
Participant 6	PIQ	108	VIQ	117
Participant 7	PIQ	111	VIQ	135
Participant 8	PIQ	88	VIQ	96

Step 1. Subtract the pairs of scores from each other in the following manner.

	VIQ		PIQ	Difference scores (D Scores)
Participant 1	112	−	109	= 3
Participant 2	117	−	110	= 7
Participant 3	99	−	105	= −6
Participant 4	128	−	114	= 14
Participant 5*	120	−	103	= 17
Participant 6*	117	−	108	= 9
Participant 7*	135	−	111	= 24
Participant 8*	96	−	88	= 8

* Note that participants 5–8 have had their scores reversed from the initial table so that all of the VIQ scores can be compared to all of the PIQ scores now that counterbalancing of the order of presentation has taken place. This means that order of presentation will no longer be a factor in determining whether VIQ is usually higher than PIQ.

Step 2. Calculate the mean of the VIQ scores ($\bar{x}_1$).

$$\bar{x}_1 = \frac{112 + 117 + 99 + 128 + 120 + 117 + 135 + 96}{8}$$

$$\bar{x}_1 = \frac{924}{8}$$

$$\bar{x}_1 = 115.5$$

Step 3. Calculate the mean of the PIQ scores ($\bar{x}_2$).

$$\bar{x}_2 = \frac{109 + 110 + 105 + 114 + 103 + 108 + 111 + 88}{8}$$

$$\bar{x}_2 = \frac{848}{8}$$

$$\bar{x}_2 = 106.0$$

Step 4. Calculate the sum of the squares of the differences between the pre-test scores and the post-test scores (ΣD^2).

$$\Sigma D^2 = 3^2 + 7^2 + (-6)^2 + 14^2 + 17^2 + 9^2 + 24^2 + 8^2$$

$$\Sigma D^2 = 9 + 49 + 36 + 196 + 289 + 81 + 16 + 64$$

$$\Sigma D^2 = 740$$

Step 5. Obtain the square of the sum of the differences between the pre-test scores and the post-test scores $(\Sigma D)^2$.

$$(\Sigma D)^2 = (3 + 7 - 6 + 14 + 17 + 9 + 24 + 8)^2$$

$$(\Sigma D)^2 = (76)^2$$

$$(\Sigma D)^2 = 5776$$

Step 6. Determine the number of pairs of scores. In the formula, N refers to the number of pairs of scores.

$$N = 8$$

Step 7. Enter the values obtained from steps 1–6 into the formula for the dependent t test.

$$t = \frac{\bar{x}_1 - \bar{x}_2}{\sqrt{\dfrac{\Sigma D^2 - \dfrac{(\Sigma D)^2}{N}}{N(N-1)}}}$$

$$t = \frac{115.5 - 106.0}{\sqrt{\dfrac{740 - \dfrac{76^2}{8}}{8(8-1)}}}$$

$$t = \frac{9.50}{0.567}$$

$$t = 16.756$$

Step 8. Compare the derived t value to the critical values of t in Appendix B. The formula for the df in a dependent t test is $df = N - 1$, where N = the numbers of scores. In the present example, there were 8 pairs so $df = 8 - 1$, or $df = 7$.

Step 9. Determine whether the H_o should be retained or rejected by comparing the derived t value to the tabled critical values of t at $p = .05$ with $df = 7$. If the derived t value exceeds the tabled critical value of t (or if it is less than the negative t critical value), then H_o is rejected. If not, then H_o is retained.
   The null and alternative hypotheses are:

$$H_o: \mu_{VIQ} = \mu_{PIQ}$$

$$H_a: \mu_{VIQ} \neq \mu_{PIQ}$$

Step 10. Write-up your findings and make a conclusion.
   Thus, the derived $t = 16.756$ does exceed the tabled critical value of $t = \pm 2.365$ at $p = .05$ with $df = 7$. Therefore, H_o is rejected, and it is concluded that the mean VIQ score (115.5) was significantly greater than mean PIQ score (106.0), $t(7) = 16.756$, $p < .001$. In terms of the research question, it appears that in this sample of patients the VIQ scores do exceed their PIQ scores, and the finding does not appear to be accounted for by order of presentation.

Determination of Effect Size

Once again, the significance of a statistical test presents only part of the picture. Effect size estimation is useful for determining the strength of the independent variable, regardless of whether H_o was retained or rejected. In the present example, the effect size determination would be:

$$r = \sqrt{\frac{t^2}{t^2 + df}}$$

$$r = \sqrt{\frac{16.756^2}{16.756^2 + 7}}$$

$$r = .99$$

By comparing this value to the table of effect sizes on page 151, we see that it translates into almost the largest effect size possible.

KEY SYMBOLS AND TERMS

Dependent *t* test – A test designed to determine the statistical difference between two means where the participants in each group are either the same or matched-pairs.

Repeated Measures Design – One design of the dependent *t* test where the participants are the same in both groups and usually are measured pre-treatment and post-treatment.

Matched *t* test – A design of the dependent *t* test where the participants are paired or matched on some critical variables that are known to be related to the dependent variable in the experiment.

Order of Presentation – Refers to the order in which levels of the independent variable are presented to the participants. If the conditions are not counterbalanced the participants may become fatigued or bored and do more poorly in the later conditions.

Regression Effect – A historically old finding in statistics where it was first noticed that the tallest people in families tended to have children shorter than they were and closer to the mean of the entire family's height (called regression to the mean). Also, people who first measure at the extreme end on a variable tend to have more moderate scores later.

1. A researcher wishes to determine whether a new medication controls seizures better than the patients' previous medications. The number of seizures are recorded on the previous and new medications for a period of one year for each patient. Are there mean differences between the numbers of seizures between the previous and new medications?

	Number of seizures	
	Previous medication	New medication
Subject 1	4	7
Subject 2	5	6
Subject 3	4	3
Subject 4	3	3
Subject 5	4	5
Subject 6	3	2
Subject 7	4	5
Subject 8	6	7
Subject 9	4	4
Subject 10	3	2
Subject 11	4	5
Subject 12	5	4

2. A researcher wishes to determine whether drinking coffee improves one's mood. The researcher hangs out at the Café Coffay and asks customers to fill out a 10-point good-mood bad-mood questionnaire (where 10 = great mood and 1 = bad mood) before and after their first cup of coffee that day. Is there any evidence that drinking coffee improves mood?

	Before mood rating	After mood rating
Participant 1	5	9
Participant 2	3	8
Participant 3	9	9
Participant 4	3	8
Participant 5	2	7
Participant 6	4	9
Participant 7	1	10
Participant 8	8	8
Participant 9	6	5
Participant 10	7	7
Participant 11	4	3
Participant 12	3	10
Participant 13	8	7
Participant 14	7	9
Participant 15	5	9

9

Analysis of Variance: One Factor Completely Randomized Design

A LIMITATION OF *t* TESTS AND A SOLUTION

While *t* tests are highly useful when we are comparing just two groups' means, a problem occurs when we try to compare three or more groups' means. For example, if we had an experiment where we were comparing the effectiveness of three different types of psychotherapies (Type A, B, and C), it would require three different *t* tests in order to make all possible mean comparisons (A vs. B, A vs. C, and B vs. C). This experimental design is a common one, and its analysis through *t* tests is called ***multiple t*s**. A problem occurs, however, because by doing each *t* test at $\alpha = .05$, we inflate the probability of committing the Type I Error (α). In multiple *t* test designs, the overall α level is obtained by the formula: overall $\alpha = 1 - (1 - \alpha)^n$ where n = the number of *t* tests performed and α is the conventional level of significance (.05). Thus, in the previous example for three *t* tests, overall $\alpha = 1 - (.95)^3$ or overall $\alpha = .14$, and as you are probably now aware, this is an unacceptable overall Type I Error rate.

THE EQUALLY UNACCEPTABLE BONFERRONI SOLUTION

A solution to this dilemma was proposed, known as the **Bonferroni correction**. It takes the number of *t* tests to be performed (k) and divides this value into the conventional level of significance ($\alpha = .05$). In our previous example, this would be $p = \alpha/k$ or $.05/3 = .017$. Thus, $p = .017$ becomes the error rate for each *t* test to be performed. This would mean that each *t* test must meet at least the $p = .017$ level of significance instead of $p = .05$ in order to be labeled statistically significant. The *t* value from the *t* distribution would be higher at $p = .017$ than at $p = .05$, thus, it would be harder to attain statistical significance. The problem with the Bonferroni correction is that by protecting against the Type I Error so strenuously, it increases the probability of Type II Error and reduces the power of our statistical analysis.

THE ACCEPTABLE SOLUTION: AN ANALYSIS OF VARIANCE

The theoretical foundations of **analysis of variance (ANOVA)** were developed by Ronald A. Fisher (1890–1962), the same Fisher who helped Gosset refine the development of the t test and its distribution. One purpose of ANOVA was to control for the automatic increase in the probability of the Type I Error when more than two groups' means are being compared. As its name suggests, ANOVA is concerned with analyzing the variance produced in multiple mean comparisons in order to determine whether genuine differences exist among the means.

THE NULL AND ALTERNATIVE HYPOTHESIS IN ANALYSIS OF VARIANCE

In ANOVA the focus is upon different types of variance inherent in a multi-group design, yet ANOVA is very much an extension of the t test for independent groups. A null hypothesis will be tested that states there is no difference among a number of group means. We will still obtain a single derived value and compare it to a distribution of values, the F distribution (named in honor of Fisher). If the derived F value exceeds the tabled critical value of F, then the null hypothesis will be rejected. The null hypothesis in ANOVA is as follows:

$$H_o: \mu_1 = \mu_2 = \ldots = \mu_k$$

where

μ_1 = the mean for Group 1

μ_k = the last group's mean

Although we will always analyze sample means, notice that the null hypothesis states that the population means are not different from each other. Thus, if there is no real effect of the independent variable upon the dependent variable, then it is as if each of these samples was taken from the same population. If the samples were all taken from the same population, then the sample means should not vary from each other. Only chance or random differences will differentiate among the group means.

Imagine if the dependent variable is significantly affected by the independent variable. In this case, one sample mean or all of them will be different from each other, and it will be as if each mean was sampled from a different population. Thus, the null hypothesis would be false. It is said, however, that in ANOVA, the null hypothesis is an **omnibus hypothesis**, and the F statistic in ANOVA is an **omnibus test**. Omnibus means 'covering many different situations at the same time.' For ANOVA, this means that if the null hypothesis is really false, we will not know which means are different from each other among the various groups. It could be that Group 1's mean is significantly lower than Group 2's mean but

Group 3's mean is not different than Group 2's mean or a plethora of other outcomes. Thus, in ANOVA the alternative hypothesis is simply stated 'H$_o$ is not true,' and H$_a$ cannot be represented symbolically.

THE BEAUTY AND ELEGANCE OF THE *F* TEST STATISTIC

The derived *F* value will be a ratio of variances. At its heart, this ratio (with a variance estimate in the numerator and a variance estimate in the denominator) is an elegant and ingenious comparison of two different forms of variance, **between-groups variance** and **within-groups variance**. The *F* ratio is as follows:

$$F = \frac{\text{between-groups variance}}{\text{within-groups variance}}$$

Let us examine the denominator type of variance first. Within-group variance is a measure of each participant's score from their group's mean. It is called variance because we will square the difference between any participant's score and the group's mean, and sum the resulting squares. For example, though each patient receives Psychotherapy A, not all will respond the exact same way. Even if Psychotherapy A works better than Psychotherapy B or C, individual differences will be inherent in any response measure to Psychotherapy A. Each groups' participants will vary from their group's mean mostly because of these individual differences. As noted earlier, statisticians have the rather bizarre name of **subject error** for this difference between any participant's score and their group mean. Subject error or within-subject variance is not systematically applied by the experimenter but randomly appears in the experiment by nature. There is also another unsystematic and random form of variance, called **experimental error**, which means that testing conditions may vary from participant to participant: Temperature may vary, lighting conditions, noise levels, etc. Since experimental error appears unsystematically and randomly, it is usually assumed that it will not affect one testing condition more than another.

The other major form of variance in an experiment, and the numerator of the *F* statistic, is the between-groups variance. This form of variance is systematic and applied by the experimenter. When the experimenter assigns three different forms of psychotherapy, the experimenter is generally assuming that there will be some mean differences among the groups. It is called between-group variance because it compares each group's mean with the grand mean of all participants in the study, and takes the sum of these squares. If there are really no differences among any of the means (H$_o$ is true), then the participants' scores will probably vary slightly from the grand mean because of nothing more than the individual differences (within-subject variance). If there are really differences among the means of the groups (H$_o$ is false), then there will be a large between-groups variance because the group means will vary more widely from the grand mean.

Remember, the between-groups variance will contain not only the variation of each group's mean from the grand mean but will also contain the variation of the individuals about their group mean (within-group variance).

THE *F* RATIO

The derived *F* ratio will consist of the between-groups variance in the numerator and the within-groups variance in the denominator. What would happen in an experiment where the null hypothesis is true? As we stated earlier, the between-groups variance actually has two sources, the variance from the differences among groups (if they really differ from one another) and within-groups variance (from individual differences). If the null hypothesis is true, then there would be little or just tiny chance contributions in the numerator for the between-groups variance. The numerator would mostly contain only within-groups variance. The denominator of the *F* ratio is another estimate of within-groups variance. If the null hypothesis is true, then we would have only within-groups variance in the numerator and within-groups variance in the denominator, thus, the resulting *F* value would be 1.00. To repeat: In ANOVA, if the null hypothesis is true, then the derived *F* value should be 1.00. To the extent that the null hypothesis is false, that is, that there are real differences among the groups' means, then the numerator of the derived *F* value will contain an estimate of the between-groups variance *and* within-groups variance. This will mean that the resulting *F* value will become greater than 1.00, and if the groups' mean are very far apart (the various levels of the independent variable really are different), then the resulting *F* value will be much larger than 1.00. A summary of these variance discussions appears in Table 9.1.

HOW CAN THERE BE TWO DIFFERENT ESTIMATES OF WITHIN-GROUPS VARIANCE?

You probably would not be surprised if you found out that a mathematician claimed to have proved that day is night and yin is yang but for the curious, here is an explanation of how there can be two different mathematical estimates of the same entity. The numerator's estimate of within-groups variance comes from a concept known as the sampling distribution. The **sampling distribution** is a theoretical distribution made up of all possible random samples from the same population. Any finite sample mean from a population of scores will probably vary slightly and sometimes more than just slightly from any other sample mean from this same population. A standard deviation derived for all of these sample means from the population mean is known as the **standard error of the mean**. If the square of the standard error of the mean (making it variance) is multiplied by the number of samples taken, the resulting value will be an estimate of the

Table 9.1

When the Null Hypothesis is True

$$F = \frac{\text{Between-Groups Variance} + \text{Within-Groups Variance}}{\text{Within-Groups Variance}}$$

Yet, Between-Groups Variance is nil
Thus,

$$F = \frac{\text{Within-Groups Variance}}{\text{Within-Groups Variance}}$$

Thus, F = 1.00

When the Null Hypothesis is False

$$F = \frac{\text{Between-Groups Variance} + \text{Within-Groups Variance}}{\text{Within-Groups Variance}}$$

And the Between-Groups variance is substantial
Thus,

$$F = \frac{\text{Between-Groups Variance} + \text{Within-Groups Variance}}{\text{Within-Groups Variance}}$$

Thus, F > 1.00

population's variance, σ^2, which also just happens to be an estimate of the within-groups variance.

The denominator's estimate of within-groups variance is a bit more straightforward. Within-groups variance in the denominator is obtained by simply taking each group's variance (the square of the standard deviation), adding them together, and dividing by the number of groups. The result (a mean of the groups' variances) is a very good estimate of the population's variance, σ^2. Interestingly, and here is the really ingenious part of ANOVA, the numerator's estimate of within-groups variance is only accurate if the null hypothesis is true and there are no differences among the groups' means. If the null hypothesis is false, then the within-groups variance also contains between-groups variance because, remember, it was derived from the standard error of the *means*. Thus, when the null hypothesis is false, and there really are differences between the groups' means, then the numerator's estimate of within-group variance is *inflated*. Since the derived F value is a ratio of these two variances, then the F value will become increasingly greater than 1.00 as the null hypothesis is increasingly more false (that is, there are greater and greater differences among the groups' means).

A friend of mine from college took an advanced statistics class with statistics and mathematics majors. The professor assigned an ANOVA problem, and the result was an F value greater than 0 but less than 1.00. 'Impossible,' shouted the students. 'Possible,' shouted my friend. He knew from empirical research that F

values less than 1.00 will occasionally occur. Since the two estimates of within-group variance are independent of each other and, although they are measuring the same value, they may vary slightly. This situation may occur if the null hypothesis is true, and if the numerator estimate is slightly less than the denominator estimate. An F value < 1.00 may also be indicative of violations of the assumptions of ANOVA, one of which is that there is homogeneity of variance about each of the groups' means.

ANOVA DESIGNS

The single most popular and the most powerful ANOVA design is called the **completely randomized ANOVA**. A completely randomized ANOVA has a single independent variable but two or more levels or conditions. These conditions will form the basis for the groups. For example, in the previous psychotherapy experiment, the independent variable is the type of psychotherapy given, and there are three levels or conditions: Psychotherapy A, Psychotherapy B, and Psychotherapy C. The design is said to be completely randomized because the participants will be randomly chosen from the population of participants, and they will be randomly assigned to the three conditions. The resulting three groups will be independent of each other, and no participant will be in any condition other than the one to which they were assigned.

The completely randomized ANOVA is a powerful one because, unlike correlation, we can imply a causative relationship between the independent and dependent variables. There is another ANOVA design, the *in situ* design, that is analyzed by the same formula as the completely randomized design but the implication of causation is weaker. As you may recall from the *in situ t* test design, the participants in the *in situ* ANOVA are not randomly assigned to the levels of the independent variable but they are pre-assigned by God or nature. For example, in one strange study, self-esteem (the dependent variable) was assessed between college students and death-row inmates. In this case, the participants were not randomly assigned to the two conditions of the independent variable (diagnosis or group membership) yet the completely randomized ANOVA formula will derive an F value the same way for this design. Only the implication of causation will be weakened.

ANOVA ASSUMPTIONS

The assumptions of an ANOVA are similar to the independent t test. First, it is assumed that the dependent variable is drawn from a population of values that is normally distributed (also called the normality assumption). Second, it is assumed that the participants have been randomly assigned to each group and the scores in each group are independent of each other. Third, it is assumed that the variances about each of the groups' means are not substantially different from each other (also called the assumption of homogeneity of variance).

It is important to note, however, that ANOVA is a robust statistical test and violations of the assumptions may still result in a correct statistical decision (to reject or retain the null hypothesis). Also, large samples (minimum of 10 to 15 participants per group) help minimize violations of the assumption of normality, and using an equal number of participants in each group helps to minimize violations of the assumption of homogeneity of variance.

PRAGMATIC OVERVIEW

The whole purpose of an ANOVA is to determine whether there are significant differences between two or more groups' means. All of the data in the experiment will be combined to form a single F value. This value will be compared to a distribution of critical F values. If the obtained F value exceeds the tabled critical F value, then the null hypothesis will be rejected, and it will be concluded that there is at least one mean significantly different from one other mean. If the obtained F value does not exceed the tabled critical value, then H_o will not be rejected, and it will be concluded that there are no significant differences between any of the means. Refer to Figure 9.1.

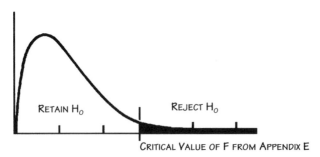

RETAIN H_o REJECT H_o

CRITICAL VALUE OF F FROM APPENDIX E

Figure 9.1

The F distribution is positively skewed with the rejection region only in the positive tail. Remember, if the null hypothesis is true, the resulting F value should be close to 1.00. If the null hypothesis is false, the resulting F value should be greater than 1.00. The critical F value will have two different types of degrees of freedom, one associated with the numerator (between-groups variance) of the F ratio and one associated with the denominator (within-groups variance).

WHAT A SIGNIFICANT ANOVA INDICATES

The interest in ANOVA is whether there are significant differences between two or more groups' means. Remember, ANOVA is an omnibus test and a significant *F* value does not reveal which groups' means are different from each other. There are a group of tests which are to be used after the null hypothesis has been rejected in ANOVA. These tests are called **multiple comparison tests**, *a posteriori* **tests** (Latin for 'what comes later') or *post hoc* **tests** (Latin for 'after this'). These tests are designed to tell which means are significantly different from each other. A significant ANOVA simply tells whether there is at least one mean different from one other mean. A multiple comparison test reveals the pattern of significant differences between the means.

A COMPUTATIONAL EXAMPLE

An experiment was done to test whether Thorazine, a major tranquilizer for the treatment of psychoses, affects driving ability. Twenty college students volunteered to serve as subjects. Five students were randomly assigned to each of four different conditions: (a) a placebo, (b) 100 mg, (c) 250 mg, (d) 500 mg. The scores obtained represent the results on a 10 point (10 = excellent, 1 = poor) driver's skills test which was performed one hour after ingestion of the drug. The data is tabled as follows:

Group			
1. Placebo	2. 100 mg	3. 250 mg	4. 500 mg
8	7	7	4
9	9	6	5
6	8	8	6
9	10	7	5
9	8	9	7

Note that each score represents a different subject and that there are a total of 20 subjects.

Step 1. Calculate Σx and Σx^2 for each of the individual groups and the overall Σx and Σx^2 for all 20 subjects.

	1	2	3	4
Σx	41	42	37	27
Σx^2	343	358	279	151

Step 2. Obtain the total or overall Σx and Σx^2.

Total $\Sigma x = 147$

Total $\Sigma x^2 = 1131$

Step 3. Calculate the correction for the mean (CM). This is done by taking the grand sum, squaring it, and dividing by the total number of observations.

$$CM = \frac{(\text{Total } \Sigma x)^2}{\text{Total Number of Scores}}$$

$$CM = \frac{(147)^2}{20}$$

$$CM = 1080.450$$

Step 4. The sum of squares total (SST) is obtained by taking the total Σx^2 and subtracting the CM.

$$SST = \text{total } \Sigma x^2 - CM$$

$$SST = 1131 - 1080.450$$

$$SST = 50.550$$

Step 5. The sum of squares between groups (SSB) is obtained by squaring the sum of each group total and dividing by the number of observations within each group, summing the quotients and subtracting the CM from the resultant value.

$$SSB = \left[\frac{(\Sigma x \text{ for Gp1})^2}{\# \text{ Scores in Gp1}} + \frac{(\Sigma x \text{ for Gp2})^2}{\# \text{ Scores in Gp2}} \right.$$
$$\left. + \frac{(\Sigma x \text{ for Gp3})^2}{\# \text{ Scores in Gp3}} + \frac{(\Sigma x \text{ for Gp4})^2}{\# \text{ Scores in Gp4}} \right] - CM$$

$$SSB = \left[\frac{(41)^2}{5} + \frac{(42)^2}{5} + \frac{(37)^2}{5} + \frac{(27)^2}{5} \right] - CM$$

$$SSB = 1108.600 - 1080.450$$

$$SSB = 28.150$$

Step 6. The error term or sum of squares within groups (SSW) is obtained by subtracting SSB from SST.

$$SSW = SST - SSB$$

$$SSW = 50.550 - 28.150$$

$$SSW = 22.400$$

Step 7. The *df* are now calculated

    df for SST = the total number of scores minus 1 = 20 −1

    df for SST = 19

    df for SSB = the total number of groups minus 1 = 4 − 1

    df for SSB = 3

    df for SSW = *df* for SST − *df* for SSB = 19 − 3

    df for SSW = 16

Step 8. The mean squares (MS) are computed by dividing the sum of squares by the appropriate *df*.

$$MSB = \frac{SSB}{df_{\text{ for SSB}}} = \frac{28.150}{3} = 9.383$$

$$MSW = \frac{SSW}{df_{\text{ for SSW}}} = \frac{22.40}{16} = 1.400$$

Step 9. The *F* value is obtained by dividing MSB by MSW.

$$F = \frac{MSB}{MSW} = \frac{9.383}{1.400} = 6.70$$

Step 10. Obtain the degrees of freedom. To enter the *F* distribution table, the *df* for SSB and SSW are needed.

Degrees of Freedom for the Numerator (df_1)

The numerator degrees of freedom or df_1 refers to the *df* for SSB. This value was obtained in Step 7. For educational purposes, it will be repeated:

    df_1 = the number of groups − 1.

Since there were four groups:

    $df_1 = 4 − 1$

    $df_1 = 3$

Degrees of Freedom for the Denominator (df_2)

The denominator degrees of freedom or df_2 refers to the *df* for SSW. This value was obtained in Step 7 but will be repeated:

df_2 or df for SSW $= df$ for SST $- df$ for SSB $= 19 - 3$

$df_2 = 16$

Step 11. Check for significance. Compare the derived F value $= 6.70$ to the tabled critical value of F at $p = .05$ with $df_1 = 3$ and $df_2 = 16$. If the derived F value exceeds the tabled critical value, then H_o is rejected. If not, then H_o is retained.
The null and alternative hypotheses are:

$H_o: \mu_1 = \mu_2 = \mu_3 = \mu_4$

where

$\mu_i = $ the mean for i^{th} group on the driver's skill test

$H_a: H_o$ is not true.

Step 12. Write up your findings and make a conclusion.
The derived $F = 6.70$ exceeds the tabled critical value of $F = 3.24$ at $p = .05$ with $df_1 = 3$ and $df_2 = 16$. Therefore, H_o is rejected, and it is concluded that at least one mean is significantly different from one other mean, $F (3, 16) = 6.70$, $p < .01$. In order to determine the pattern of mean differences, a *post hoc* test is needed. In terms of the research question, it does appear that Thorazine does significantly affect driving ability.

Step 13. Summarize the results of the ANOVA in a table as follows:

Source	Sum of squares	df	Mean squares	F	p
Between Groups	28.15 (SSB)	3	9.38 (MSB)	6.70	<.01
Within Groups	22.40 (SSW)	16	1.40 (MSW)		
Total	50.55 (SST)	19			

DETERMINING EFFECT SIZE IN ANOVA

Although we found significance in this example, we do not know the strength of the relationship between the independent variable (Thorazine levels) and the dependent variable (driving skills test). Just as in the t test, significance is only part of the picture. For ANOVA one popular measure of the magnitude of the effect of the independent variable upon the dependent variable is omega-squared (ω^2). Omega-squared is obtained through the following formula:

$$\omega^2 = \frac{SSB - (k - 1)MSW}{SST + MSW}$$

where k = the number of levels of the independent variable

In the previous example,

$$\omega^2 = \frac{28.15 - (4-1)1.40}{50.55 + 1.40}$$

$$\omega^2 = \frac{23.95}{51.95}$$

$$\omega^2 = .46$$

According to omega-squared interpretation guidelines,

$\omega^2 > .15$ = large effect

$\omega^2 > .06$ = medium effect

$\omega^2 > .01$ = small effect

Thus, in our example $\omega^2 = .46$, which is indicative of a large effect size, ω^2 may also be interpreted as the percentage of variance in the scores accounted for by the independent variable. ω^2 is a measure of how much variation is accounted for by the treatment and, by default, how much of the total variation in scores is accounted for by random factors. In this example, 46% of the total variance in the scores can be accounted for by the treatment with Thorazine and 54% of the total variance is accounted for by unknown or random factors.

HISTORY TRIVIA

In 1908, William Sealy Gosset (1876–1937) published a table of ratios of differences between sample means and population means. He published under the pseudonym 'Student' since the Guinness Brewery for whom he worked had strict regulations about publications by its employees. The article appeared in *Biometrika*, which was Karl Pearson's journal devoted to mathematical and statistical studies of the life sciences. Gosset called the ratios z's, and found that they did not fit the normal distribution. He entitled the paper, 'The Probable Error of a Mean.' He also published a second paper in *Biometrika* that same year called, 'Probable Error of a Correlation Coefficient.'

Ronald Aylmer Fisher (1890–1962) received his undergraduate degree in astronomy in England. Subsequently, he worked for an experimental agricultural station for 14 years analyzing, among other data, wheat crop yields and weather patterns. In 1912, Gosset was 36 years old and Fisher was 22. Fisher had just graduated from Cambridge University, and in his first paper he derived an estimate for the variance of a normal sample which differed from Gosset's. Fisher used a divisor of *n* where Gosset used *n* − 1. Fisher's college tutor encouraged him to write to Gosset and he did so. Thus began a friendship and an exchange of letters until Gosset's death in 1937.

Fisher's first letter to Gosset apparently gave Fisher another idea: to represent geometrically the configuration of a sample in n-dimensional space. This in turn led him to the notion of degrees of freedom and to the correct divisor for the formula to estimate variance $(n - 1)$. Finally, it led Fisher to a mathematical proof of Gosset's z distribution. Fisher sent the proof to Gosset, and Gosset sent it to Pearson with a recommendation to publish it. About the paper, Gosset wrote, 'I couldn't understand his stuff . . . I don't feel at home in more than three dimensions even if I could understand it otherwise.' But Gosset concluded, 'It's so nice and mathematical that it might appeal to some people.' Thus began their collaboration which would lead to a retabulation of Gosset's zs into ts, and it was to be known as Student's t distribution. Fisher went on to solve the problem posed in Gosset's second 1908 paper by deriving the general sampling distribution of the correlation coefficient. Upon receipt of Fisher's paper, Gosset, with his typical humility and generosity, wrote, 'When I first saw it, I nearly wrote to thank you for the kind way in which you referred to my unscientific efforts.'

When Gosset became aware that Fisher was looking for a job and that Fisher had been doing statistical work on orchards, Gosset told Fisher that John Russell of the Rothamstead Agricultural Experiment Station was looking to hire a statistician. Eight months later Gosset was notified by Fisher that he had gotten the appointment. In Fisher's letter he asked for advice on what calculator he should buy and advice on home brewing. With respect to the latter request, Gosset wrote, '. . . less trouble to buy Guinness . . . let us do it for you . . .'

At Rothamstead, Fisher would create two of his greatest works, *Statistical Methods for Research Workers* published in 1925, and *The Design of Experiments* published in 1935. Fisher came to America in 1931 and taught at Iowa and Minnesota. At Iowa he was to influence George W. Snedecor. Snedecor published his own influential statistical text in 1937, and analysis of variance would become common for statistical analyses in agriculture. Beyond Fisher's tremendous contributions to hypothesis testing with small samples and their test distributions, he would ultimately contribute two of statistics' most popular tests, the t test and analysis of variance.

An instructor went to teach at University College during the years of Karl Pearson. What was her reaction to these famous figures? Of Karl Pearson, she wrote, 'Endured K.P.' Of Neyman, she wrote, 'I was baby-sitting for Neyman, explaining to the students what the hell he was up to.' She wrote, 'Went fly fishing with Gosset. A nice man . . . (he) didn't have a jealous bone in his body.'

Multiple *ts* – a design where mean differences among three or more groups are assessed by more than one *t* test. Using multiple *ts* has the effect of inflating the overall α level (the probability of committing the Type I Error) to unacceptably high levels ($p > .05$).

Bonferroni Correction – A solution created to correct for a high overall α level when using multiple *t* tests which divides overall α level by the number of *t* tests to be performed. The Bonferroni correction does reduce the probability of the Type I Error by tremendously increasing the probability of the Type II Error.

Analysis of Variance (ANOVA) – A popular, powerful, and robust statistical test that assesses the differences between two or more groups' means by analyzing a ratio of variances between groups and within groups.

Omnibus Hypothesis and Omnibus Test – The word 'omnibus' means covering many situations at once. ANOVA has omnibus hypotheses because the actual pattern of mean differences is not specified. ANOVA is an omnibus test because any pattern of mean differences can cause the rejection of the null hypothesis.

Between-Groups Variance – A measure of the sum of the squared differences between a series of groups' means and the population mean (or grand mean). Between-groups variance is a measure of how much the independent variable affects the dependent variable in ANOVA.

Within-Groups Variance – A measure of the sum of the differences of each participant's score from the group's mean. Within-groups variance is a measure of how much individual differences affect the dependent variable even when all of the participants are treated alike.

Subject Error – A curious name for the difference between a participant's score in a group and the group's mean.

Experimental Error – The statistical name for the random and unsystematic effects that influence an experiment but are not due to the treatment or independent variable. The largest source of experimental error is subject error and may also include variations in testing conditions like temperature, lighting, experimenter's changing enthusiasm for the experiment, participant's variation in interest, boredom, etc.

Sampling Distribution – A theoretical distribution made up of all possible finite random samples from the same population.

Standard Error of the Mean – The standard deviation of the sampling distribution's means.

Completely Randomized ANOVA – The most powerful ANOVA design where participants are randomly assigned to one of the two or more groups. Causation may be implied from a completely randomized ANOVA design.

Multiple Comparison Tests – Also known as *a posteriori* or *post hoc* tests. Multiple comparison tests determine the pattern of mean differences after the null hypothesis has been rejected in an ANOVA design.

1. What is the purpose of an ANOVA?

2. Name some types of experimental error.

3. How can you control for experimental error caused by subject differences?

4. How do the *F* values differ when the null hypothesis is true and when the null hypothesis is false?

5. What type of procedure is used when the ANOVA is significant? Why is this done?

6. What are the advantages of a completely randomized design?

7. Find the critical *F* values for the following:

   a) $= .05$, df between $= 2$, *df* within $= 20$

   b) $= .01$, df between $= 5$, *df* within $= 50$

   c) $= .05$, df between $= 12$, *df* within $= 12$

   d) $= .01$, df between $= 9$, *df* within $= 16$.

8. A researcher wishes to determine which of three therapy methods reduces arthritic patients' pain. The researcher randomly assigns 18 subjects to one of three conditions, (a) Zen relaxation, (b) hypnosis, (c) control. The subjects were then either trained in Zen, were hypnotized, or were given no special help (control group). Later, all three groups were asked to measure the number of times in one week that their pain became excruciating. The number of pain episodes for each patient follows:

Zen Group	Hypnosis Group	Control Group
11	10	16
12	8	15
10	9	14
10	7	12
9	5	16
9	3	20

   a.   Perform an analysis of variance.

   b.   Describe, fully, in your own words, completely, what the experiment and the subsequent analyses mean!

9. A researcher wishes to determine whether a new drug called 'Unanx' reduces social anxiety in schizotypal people. Forty schizotypal people were randomly assigned to one of 5 conditions (a) placebo condition (no drug), (b) 5 milligrams, (c) 10 milligrams, (d) 15 milligrams, (e) 20 milligrams. One hour after ingestion of the drug they were given a behavioral test of social anxiety by observing their behavior in a crowded room. They were rated by a graduate student on a 1 to 10 scale (where 10 = no social anxiety and 1 = massive social anxiety). The social anxiety scores are as follows:

Placebo	5mgs	10mgs	15mgs	20mgs
5	6	6	8	9
6	4	5	7	10
5	6	4	9	9
4	5	7	10	10
7	3	8	10	10
5	7	5	9	10
6	6	7	8	9
5	6	6	8	9

   a. Does the Unanx reduce social anxiety?

   b. What do the results mean? What can be concluded?

   c. What confounding variables could have affected the outcome of this experiment?

10. A dietician wishes to determine which of four diets appears to help people lose weight. Is there any evidence of a superior diet? Twenty people were randomly assigned to one of four diets for six weeks. The numbers represent weight loss in pounds for each participant.

Diet 1 = lots of fruit

Diet 2 = protein oriented

Diet 3 = calorie counting

Diet 4 = Zen meditation, and Group Therapy

Fruit	Protein	Count	Zen
14	11	10	25
19	15	5	18
12	9	6	24
9	19	11	29
14	12	4	17

10 After a Significan Analysis of Variance: Multiple Comparison Tests

The purpose of ANOVA was to determine whether significant differences existed among two or more groups' means without inflating the probability of the Type I Error. One limitation of ANOVA is that, if the null hypothesis is rejected, ANOVA is an omnibus test with a vague alternative hypothesis (H_o is not true). ANOVA does not tell us what the exact pattern of mean differences is. **Multiple comparison tests**, also called *a posteriori tests* or *post hoc tests*, were designed to be used after a significant ANOVA where the null has been rejected. Multiple comparison tests help to determine the pattern of significant differences among the means, and they also keep the Type I Error rate at an acceptable level ($p = .05$ or less) despite whether there are two or more comparisons to be made among the groups' means.

There are many different multiple comparison tests such as Tukey's, Newman-Keuls, Scheffe's, Duncan's, etc. Some of these are considered to be more conservative in controlling the Type I Error and others are considered more liberal. This book will present only one: **Tukey's HSD test**. It is a popular multiple comparison test, and it is considered to be neither too conservative nor too liberal.

Historically, one of the first multiple comparison tests to be used after a significant ANOVA was the **Least Significant Difference test (LSD test)**. While college statistics students in the 1960s snickered at the initials, it quickly came to be known that the test was far too liberal, that is, the probability of the Type I Error was increased beyond an acceptable level (*e.g.*, $p > .05$). Contemporary American statistician John Tukey created a better test that came to be known as the **Honestly Significant Difference test** or **Tukey's HSD test** or now simply as **Tukey's test**.

The basic assumptions made for *t* tests are also required for Tukey's test (*e.g.*, normal distributions, homogeneity of variance, and approximately equal numbers of subjects).

CONCEPTUAL OVERVIEW OF TUKEY'S TEST

Each mean in the rejected null hypothesis will be compared to every other mean. If there are three group means, then there will be three comparisons (Group 1 versus Group 2, Group 1 versus Group 3, and Group 2 versus Group 3). If there are four group means, then there will be six comparisons, and if there are five group means, then there will be 10 comparisons. The value of the difference between two means will be compared to a critical value, known as Tukey's HSD value. The absolute value of the difference between the two means must exceed Tukey's HSD value in order to be statistically significant.

COMPUTATION OF TUKEY'S HSD TEST

Step 1. Determine all possible mean comparisons that can be made from the rejected null hypothesis. For example, if the rejected null hypothesis had three means then there are three possible mean comparisons: $\bar{x}_1$ minus $\bar{x}_2$, $\bar{x}_1$ minus $\bar{x}_3$, and $\bar{x}_2$ minus $\bar{x}_3$. If the rejected null hypothesis contained four means, then there are six mean difference comparisons.

For example, in the previous chapter (one factor-completely randomized design) a significant F value was obtained. The null hypothesis contained four means (placebo group, 100 mg group, 250 mg group, and 500 mg group). There are six possible mean comparisons which are as follows:

$$\bar{x}_{placebo} \text{ versus } \bar{x}_{100mg}$$

$$\bar{x}_{placebo} \text{ versus } \bar{x}_{250mg}$$

$$\bar{x}_{placebo} \text{ versus } \bar{x}_{500mg}$$

$$\bar{x}_{100mg} \text{ versus } \bar{x}_{250mg}$$

$$\bar{x}_{100mg} \text{ versus } \bar{x}_{500mg}$$

$$\bar{x}_{250mg} \text{ versus } \bar{x}_{500mg}$$

Step 2. Determine the absolute value of the difference between each comparison.

Comparison		Means		Absolute value of difference
$\bar{x}_{placebo}$ minus $\bar{x}_{100mg}$	=	8.2 – 8.4	=	.2
$\bar{x}_{placebo}$ minus $\bar{x}_{250mg}$	=	8.2 – 7.4	=	.8
$\bar{x}_{placebo}$ minus $\bar{x}_{500mg}$	=	8.2 – 5.4	=	2.8
$\bar{x}_{100mg}$ minus $\bar{x}_{250mg}$	=	8.4 – 7.4	=	1.0
$\bar{x}_{100mg}$ minus $\bar{x}_{500mg}$	=	8.4 – 5.4	=	3.0
$\bar{x}_{250mg}$ minus $\bar{x}_{500mg}$	=	7.2 – 5.4	=	2.0

Step 3. Determine Tukey's HSD critical value according to the following formula.

$$HSD = q \sqrt{\frac{MSW}{N}}$$

First, obtain the value of q from Appendix F (Tukey's HSD q Table of Critical Values). In order to enter this table, we will be required to know three values:

1. Significance level or p level = .05 or .01 (we will use .05).

2. r = the number of different means that will be compared (hint: look at H_o and count the means).

3. df_{MS} or df_{error} = the degrees of freedom associated with the MS error term or the significant F value.

Our Tukey's test will be conducted at p = .05. In our previous example, we had four unique means in the null hypothesis, so r = 4, and df_{error} = 16 (df for MSW). Thus, the q value is 4.05.

Next, determine the value of MSW (or MS within groups). This value is also called the MS error term. The value of MSW was tabled in step 13 of Chapter 9 as the Mean Squares for the Within Groups source. It was also the value of the denominator in the F value. In the present example, MSW = 1.400.

Finally, N = number of scores upon which **each mean** was based (if Ns are not equal, use the harmonic mean described later in this chapter).

Thus,

$$\text{HSD} = 4.05 \sqrt{\frac{1.400}{5}}$$

$$\text{HSD} = (4.05)(.5292)$$

$$\text{HSD} = 2.14$$

Step 4. The obtained HSD value is a critical value. **The differences between the means in Step 2 must exceed this critical value in order to be significantly different at p < .05.** Thus, in order to be significantly different the absolute value of a mean pair difference must exceed 2.14.

Now, attach an asterisk to those mean pairs which exceed the critical HSD value.

Comparison		Means		Absolute value of difference
$\bar{X}_{placebo}$ minus $\bar{X}_{100mg}$	=	8.2 – 8.4	=	.2
$\bar{X}_{placebo}$ minus $\bar{X}_{250mg}$	=	8.2 – 7.4	=	.8
$\bar{X}_{placebo}$ minus $\bar{X}_{500mg}$	=	8.2 – 5.4	=	2.8*
$\bar{X}_{100mg}$ minus $\bar{X}_{250mg}$	=	8.4 – 7.4	=	1.0
$\bar{X}_{100mg}$ minus $\bar{X}_{500mg}$	=	8.4 – 5.4	=	3.0*
$\bar{X}_{250mg}$ minus $\bar{X}_{500mg}$	=	7.2 – 5.4	=	2.0
* p < .05				

DETERMINING WHAT IT ALL MEANS

More often than not, the overall pattern of significant and nonsignificant differences among the means will be understandable. It is also important to remember that nonsignificant differences may have just as much meaning as significant ones. Let us start with the first pair of means. From that pair (the placebo and the 100 mg group), it can be determined that the mean driving score for the placebo group (8.2) is not significantly different than the mean score for the 100 mg group (8.4). What does that indicate? Well, our preliminary conclusion would be that 100 mg of Thorazine does not significantly impair (nor improve) driving ability since there was no significant difference between those two means. You may have also noticed that *mathematically* the mean of the 100 mg group (8.4) was higher than the placebo group's mean (8.2). However, **do not interpret this difference**! It is extremely important to remember that Tukey's test has shown that this difference is a nonsignificant one, so ignore any mathematical differences. If there is no statistically significant difference between two means, then any mathematical differences between them are attributed to chance, and these chance differences should be completely ignored.

The next pair of means (placebo and 250 mg groups) are also not significantly different from one another. This indicates that the mean driving score for the placebo group (8.2) is not significantly higher or lower than the 250 mg group's mean (7.4). Again, the interpretation is the same as before. It appears that 250 mg of Thorazine also does not impair driving ability. Our preliminary conclusion would be that up to 250 mg of Thorazine does not impair driving ability.

Once again, it is important to warn you that a mathematical difference between the two means is apparent. However, do not interpret this mathematical difference. Since it is not statistically significant according to Tukey's test, then the differences between the two means are attributed to chance.

The third pair of means (placebo and 500 mg groups) is significantly different from each other. This indicates that the placebo mean driving score (8.2) is significantly higher than the 500 mg group's driving score mean (5.4). Thus, our conclusion would be that 500 mg of Thorazine does appear to impair driving ability, and while up to 250 mg of Thorazine does not impair driving ability, 500 mg does impair driving ability.

We still have three more pairs of means to interpret yet notice that a meaningful overall pattern has already emerged. 100 mg and 250 mg of Thorazine do not affect driving ability but 500 mg has a negative effect on driving ability. We discovered this pattern by comparing all three Thorazine groups (100 mg, 250 mg, and 500 mg) to the placebo group's mean. The last three pairs of differences do not provide any more appreciable meaning to our overall pattern. The 100 mg group's driving ability is not different from the 250 mg group's, the 100 mg group's driving ability is significantly less than the 500 mg group's driving ability, and the 250 mg group's driving ability is not significantly different from the 500 mg group's driving ability.

ON THE IMPROTANCE OF NONSIGNIFICANT MEAN DIFFERENCES

It is very important to note that most researchers, casual readers, journal editors, and just about everybody else in the world are biased towards significant differences. It is important to remember that **nonsignificant differences may be just as important as significant differences**. In the previous example, the first and second mean pairs are not significantly different but they contain very important information. The first pair reveals that the mean driving score for the placebo group (8.2) is not significantly different than the mean score for 100 mg group (8.4). This means that Thorazine does not seem to impair driving ability when given in a 100 mg dose and that may be very important information to many people. In addition, there was no significant difference between the mean driving score for the placebo group (8.2) and the mean score for the 250 mg group (7.4). This means that Thorazine may not impair driving ability in a 250 mg dose. Again, this may be important information although it was not statistically significant. However, remember the word *significance* or *nonsignificance* in a statistical context means *not likely due to chance* or *likely due to chance*, and it does not represent a value judgment about the value of the results. So remember, when interpreting the results of a multiple comparison test, pay close attention to the significant *and* the nonsignificant mean pairs.

TUKEY'S WITH UNEQUAL NS

If the means are based on unequal Ns, the N in Tukey's formula is based on the **harmonic mean**. The harmonic mean is obtained by the following formula:

$$\tilde{N} = \frac{k}{\dfrac{1}{N_1} + \dfrac{1}{N_2} + \ldots + \dfrac{1}{N_k}}$$

where k = the number of means

N_1 = the number of scores in Group 1

N_k = the number of scores in the last group

$\tilde{N}$ = the harmonic mean

Multiple Comparison Test – is a test to be used only after a significant ANOVA. It determines which pairs of means are significantly different from each other.

A Posteriori **Test** – is an alternative name for a multiple comparison test.

Post Hoc **Test** – is an alternative name for a multiple comparison test.

Least Significant Difference Test – historically, one of the first multiple comparison tests that is presently considered too liberal (*i.e.*, *p* level may exceed .05).

Tukey's HSD Test – a popular multiple comparison test, considered neither too liberal nor too conservative which maintains the Type I Error rate regardless of the number of means to be compared.

Harmonic Mean – an alternative mathematical concept and formula to the standard mean or average formula. It is to be used in Tukey's HSD test when there are an unequal number of scores in each group.

1. When are multiple comparison tests used?

2. If the rejected null hypothesis had five means, what are the possible mean comparisons?

3. What are the basic assumptions required for Tukey's post hoc test?

4. In what ways might nonsignificant differences be important?

5. What value must the differences between the means exceed to be significant?

6. What should be done if the means are based on unequal Ns?

7. What is the difference between the LSD test and HSD test?

TRUE/FALSE

8. If the F value in an ANOVA is not significant, a multiple comparison test can be still used to test for differences among the means.

9. The HSD becomes a critical value which the actual mean differences must exceed.

10. Nonsignificant differences are not as meaningful as significant differences.

11. A neuropsychologist wishes to determine whether fine motor performance improves for learning disabled children under various drug conditions. Forty children were randomly assigned to one of four conditions: Placebo, Cylert (a stimulant), Caffeine (a stimulant), and Tofranil (an antidepressant). The participants were tested for fine

Placebo	Cylert	Caffeine	Tofranil
30	37	36	25
35	41	39	28
31	43	45	22
38	44	42	26
36	40	42	23
32	45	41	26
34	38	39	30
34	39	40	25
31	42	43	22
37	37	41	24

motor movement by the Finger Tapping Test, and the scores represent the number of taps in a 12 second period. A larger score indicates better fine motor performance. Do an appropriate statistical analysis. What can be concluded? What are appropriate criticisms?

11 Analysis of Variance: One Factor Repeated Measures Design

THE REPEATED MEASURES DESIGN

A one factor **ANOVA-repeated measures design** is similar to a repeated measures *t* test. The same participant serves under all levels of the single factor. In the repeated measures *t* test, the same participant served in both groups. In the one factor repeated measures ANOVA, the same participant serves under all levels of the single factor. The interpretation of the repeated measures ANOVA is the same as the completely randomized ANOVA. The interest in ANOVA is whether there are significant differences between two or more groups' means. Remember that a significant ANOVA does not reveal which groups' means are different from each other. A multiple comparison test or *post hoc* test is designed to tell which means are significantly different from each other. A significant ANOVA simply tells whether there is at least one mean different from one other mean. The multiple comparison test reveals the pattern of significant differences between the means.

One variant of the repeated measures *t* test is that there could be pairs of participants who were matched on some critical variable or variables. The same is true of the repeated measures ANOVA. The variant of the repeated measures ANOVA is called the **randomized block design**. Instead of the same participant serving under all levels of a factor, a block of participants who are all alike are randomly assigned to each level of the single factor. The randomized block design is probably more common in animal experiments where the various blocks may be litters of rats or armadillos (a litter of armadillos consists of identical quadruplets which makes them ideal participants for a one-factor randomized block ANOVA).

In some experiments the repeated measures design may not be used. Experiments where the participants must be naive in each condition will preclude the use of a repeated measures ANOVA. Other examples are in drug studies where one level of a drug will affect other levels, or in designs where the levels of a factor are fixed such as an analysis of three different types of personality disorders, or a cross-sectional study of four different age groups.

One advantage of the repeated measures design is that it utilizes fewer participants than the completely randomized design. Another advantage is that the statistical power of the test is increased because the variability of one participant across all levels of the factor is usually less than the variability between different participants across levels of the factor.

ASSUMPTIONS OF THE ONE FACTOR REPEATED MEASURES ANOVA

The repeated measures ANOVA assumptions are the same as in the one factor completely randomized ANOVA with one addition. Thus, it is assumed that the distributions of the populations of scores are normal, the scores are independent of one another, and the variances of the populations are homogeneous. The additional assumption concerns the homogeneity of variance of the difference scores between each level of the independent variable for the participants. A difference score is derived for each participant by subtracting their score on one level from their score on another level. These difference scores across all participants are assumed to have equal (or homogeneous) variances in the population of difference scores.

COMPUTATIONAL EXAMPLE

Eight participants with a diagnosis of Alzheimer's disease were tested by a neuropsychologist at the end of each of 4 consecutive years. The scores were examined in order to determine whether cognitive deficits increase over time. The patients were rated on a scale from 0, which represented no deficits, to 11, which indicated maximum deficits. The data is tabled as follows:

| Participants | Years | | | |
	1st	2nd	3rd	4th
P_1	7	8	8	7
P_2	9	8	10	11
P_3	5	6	8	7
P_4	10	9	10	11
P_5	4	5	7	8
P_6	5	4	5	5
P_7	6	5	6	7
P_8	8	9	9	10

Step 1. Obtain Σx and Σx^2 for the four different levels of the factor and for overall.

| | Years | | | | |
	1	2	3	4	Overall
$\Sigma x =$	54	54	63	66	$\Sigma x = 237$
$\Sigma x^2 =$	396	392	519	578	$\Sigma x^2 = 1885$

Step 2. The correction for the mean (CM) is obtained by squaring the total Σx and dividing by N (the total number of observations).

$$CM = \frac{(\text{total } \Sigma x)^2}{N} = \frac{237^2}{32} = 1755.281$$

Step 3. The sum of squares total (SST) is obtained by taking the total Σx^2 and subtracting the CM.

$SST = \text{total } \Sigma x^2 - CM$

$SST = 1885 - 1755.281$

$SST = 129.719$

Step 4. Obtain Σx for each participant across the four levels of the treatment.

$P_1 = (7 + 8 + 8 + 7) = 30$

$P_2 = (9 + 8 + 10 + 11) = 38$

$P_3 = 26$

$P_4 = 40$

$P_5 = 24$

$P_6 = 19$

$P_7 = 24$

$P_8 = 36$

Next, square each of these values, add these results together, and divide by the number of scores upon which each sum was based.

$$\frac{30^2 + 38^2 + 26^2 + 40^2 + 24^2 + 19^2 + 24^2 + 36^2}{4} = 1857.250$$

Next, the sum of squares for participants (SSS) is obtained by subtracting the CM from the above obtained value.

$SSS = 1857.250 - CM$

$SSS = 1857.250 - 1755.28$

$SSS = 101.969$

Step 5. In order to obtain the sum of squares for treatments (SSB), obtain Σx for each level of the factor across all the participants in a group, which was done in step 1, square each of these four values and add them together and divide by the number of scores upon which each sum was based.

$$\frac{(\Sigma x_1)^2 + (\Sigma x_2)^2 + (\Sigma x_3)^2 + (\Sigma x_4)^2}{N} = \frac{54^2 + 54^2 + 63^2 + 66^2}{8}$$

To obtain SSB, take the resulting value and subtract the CM.

SSB = 1769.625 – CM

SSB = 1769.625 – 1755.281

SSB = 14.344

Step 6. In order to obtain the error term sum of squares (SSW), take SST and subtract SSS and SSB.

SSW = SST – SSS – SSB

SSW = 129.719 – 101.969 – 14.344

SSW = 13.406

Step 7. The degrees of freedom *(df)* are now calculated.

df for SST = total number of scores – 1

= 32 – 1

= 31

df for SSS = total number of participants – 1

= 8 – 1

= 7

df for SSB = total number of treatment levels – 1

= 4 – 1

= 3

df for SST = *df* for SSS – *df* for SSB

= 31 – 7 – 3

= 21

Step 8. The mean squares (MS) are now calculated, and are obtained by dividing the sum of squares by the appropriate *df*.

$$MSB = \frac{SSB}{df_{SSB}} = \frac{14.344}{3} = 4.781$$

This value is not needed. The interest in this design is
MSS = not in the difference among participants but
in the stability in cognition across time.

$$MSW = \frac{SSW}{df_{SSW}} = \frac{13.406}{21} = 0.638$$

Step 9. The F value for testing significance is obtained by dividing MSB by MSW.

$$F = \frac{MSB}{MSW} = \frac{4.781}{0.638}$$

$$F = 7.49$$

Step 10. In order to check the obtained F value for significance, refer to the F distribution. To enter the table, the df for SSB and SSW are needed. The df for the numerator or df_1, refers to the df for SSB (or number of levels of the main factor minus 1).

The df for the denominator or df_2, refers to the df for SSW.

In this example $df_1 = 3$ and $df_2 = 21$. The critical value at $p = .05$ is 3.07. Our obtained F value exceeds this value; therefore, it is significant, $F(3,21) = 7.49$, $p < .05$.

The null and alternative hypotheses are:

$$H_0: \mu_1 = \mu_2 = \mu_3 = \mu_4$$

where

μ_i = the mean for the ith group on the cognitive deficits scale.

H_a: H_0 is not true.

Since the F value was significant, we reject the null hypothesis.

Step 11. The analysis is tabled as follows:

Source	Sum of squares	df	Mean squares	F	p
Between Groups	14.34	3	4.78	7.49	<.01
Subjects	101.97	7			
Within Groups (error term)	13.40	21	.64		
Total (SST)	129.72	31			

Step 12. Write up your findings and make a conclusion.

The derived value of $F = 7.49$ exceeds the critical value of $F = 3.07$ at $p = .05$ with $df_1 = 3$ and $df_2 = 21$. Therefore, H_0 is rejected, and it is concluded that at least one mean is significantly different from one other mean, $F(3,21) = 7.49$, $p < .01$.

Remember, a significant analysis of variance does not reveal which means are significantly different from each other. In order to determine the pattern of mean differences a *post hoc* or multiple comparison test must be employed. In terms of the research question, it appears that cognitive deficits do appear to change over a four year testing period.

DETERMINING EFFECT SIZE IN ANOVA

Just as in the one factor completely randomized ANOVA, we still do not know the strength of the relationship between the independent variable and the dependent variable after an ANOVA. If there is a significant F value, we know that at least one mean is significantly different from one other mean, but only an effect size analysis will give an estimate of how much the independent variable affects the dependent variable. Once again, the magnitude of the effect of the independent variable upon the dependent variable will be calculated by the omega-squared (ω^2) statistic. Omega-squared is obtained through the same formula presented in Chapter 9 and is as follows:

$$\omega^2 = \frac{SSB - (k - 1)MSW}{SST + MSW}$$

where k = # of levels of the independent variable

In the previous example,

$$\omega^2 = \frac{14.34 - (4 - 1)0.64}{129.72 + 0.64}$$

$$\omega^2 = \frac{12.42}{130.36}$$

$$\omega^2 = .095$$

According to omega-squared interpretation guidelines,

$\omega^2 > .15 =$ large effect

$\omega^2 > .06 =$ medium effect

$\omega^2 > .01 =$ small effect

Thus, in our example $\omega^2 = .10$, which is indicative of somewhere between a medium to large effect size. ω^2 may also be interpreted as the percentage of variance in the scores accounted for by the independent variable. ω^2 is a measure of how much variation is accounted for by the treatment, and how much of the total variation in scores is accounted for by random factors. In this example, 10% of the total variance in the scores can be accounted for by the time periods factor and 90% of the total variance is accounted for by unknown or random factors.

ANOVA-Repeated Measures Design – is similar to the dependent t test design in that the same participants are used for each level of the independent variable. The research interest is whether the means for each level of the independent variable are significantly different from each other.

Randomized Block Design – a form of repeated measures design in ANOVA that uses groups of similar participants (like triplets or quadruplets) so that, although the participants are different for each level of the independent variable, the repeated measures ANOVA formula is still used to analyze the data.

1. A psychologist wishes to determine whether children's IQs change while they attend elementary school. Eight children are tested with the Wechsler Intelligence Scale for Children after each year of school. Is there any evidence that IQ changes across the grade levels? Discuss your findings.

| | Grade | | | | |
	1	2	3	4	5
Child 1	101	100	102	100	105
Child 2	115	110	120	116	117
Child 3	145	138	147	132	148
Child 4	97	99	101	88	90
Child 5	105	110	109	103	105
Child 6	112	122	108	119	110
Child 7	105	99	98	100	107
Child 8	135	134	130	135	130

2. A group of morbidly obese patients (more than 100 lbs. over ideal weight) were hypnotized to increase their will power to lose weight. The sessions were conducted by a psychologist once a month for four months. The patients' weights were measured at the end of each month for the four month period. Is there evidence that the patients lost weight? Perform an analysis of variance and discuss your findings. Do a post hoc test if necessary. What is a limitation or criticism of the study?

	Month 1	Month 2	Month 3	Month 4
Patient 1	245	225	219	202
Patient 2	340	311	305	284
Patient 3	298	280	277	244
Patient 4	355	328	309	285
Patient 5	321	322	330	325
Patient 6	414	387	375	340
Patient 7	309	285	277	249

12 Analysis of Variance: Two Factor Completely Randomized Design

FACTORIAL DESIGNS

While one factor ANOVA designs are powerful and elegant, the world is a complex place. It is rare that we can attribute behavior to a single variable. For example, the roots of war, poverty, aggression, crime, suicide, or a plethora of other societal ills can never be attributed to a single causal agent. In recognition of this fact, statisticians have devised multi-factor experimental designs and analyses to account for the confluence of more than one variable upon behavior. In ANOVA these designs are referred to as factorial designs. A **factorial design** evaluates the effects of two or more independent variables (called factors) simultaneously upon a single dependent variable. The simplest factorial ANOVA design is the **completely randomized factorial design** where there are two or more factors (or treatments) which each have two or more levels. The phrase 'completely randomized' indicates that the participants have been randomly assigned to one of the unique levels of the factors. If there are x levels of one factor and y levels of another factor, then there are x times y number of factor combinations and each participant can serve in only one unique condition (also called a cell). Factorial designs are sometimes referred to as a 2 × 3 (pronounced two by three) factorial or 2 × 3 × 3 factorial design. The individual numbers refer to the number of levels of each factor and the number of numbers refers to the number of factors in the experiment. Thus, there are 3 independent variables in a 2 × 3 × 3 factorial ANOVA, and there are 18 different treatment conditions. In my master's thesis, I had a 2 × 2 × 2 × 2 × 3 ANOVA design. Therefore, I was testing five independent variables, four of them had two levels and one had three levels.

THE MOST IMPORTANT FEATURE OF A FACTORIAL DESIGN: THE INTERACTION

The most important feature of a factorial design is the **interaction** between the two independent variables upon the dependent variable. An interaction between the two factors allows the experimenter to determine what the effect is of both independent variables simultaneously upon the dependent variable. If we performed an experiment with one independent variable and analyzed it with a one factor ANOVA and then performed a second one factor ANOVA on a second independent variable, we would still not know the effects of the two variables together. This interaction between the two independent variables is one of the powerful advantages of factorial designs in ANOVA, and it explains their popularity in the scientific literature. Each of the two independent variables is called a **main effect**. Thus, there are two main effects and one interaction in a two factor ANOVA.

There are also said to be **simple effects** in a factorial ANOVA. In this case, the experimenter is interested in all levels of one of the independent variables under only one level of the other independent variable. The interaction should be significant for the simple effects to be of interest to an experimenter since, if there was no significant interaction, the main effects alone could have predicted the outcome of the experiment.

There is a cost of a factorial ANOVA design, and it is that we will sacrifice experimental simplicity, ease of interpretation, and ease of computation. A factorial design is also usually a large experiment and, if many treatment levels are involved, the number of participants can become cumbersomely large. Overall, however, the benefits of the two factor ANOVA and its interaction far outweigh these costs.

FIXED AND RANDOM EFFECTS AND *IN SITU* DESIGNS

The independent variable in ANOVA designs may vary in nature. It is said to be a **fixed effect** if the experimenter is only interested in those particular levels of the independent variable and is not interested in generalizing to other levels not included in the experiment. For example, if gender was chosen as the independent variable, then it would be a fixed effect because there are only (typically) two levels of gender. In this case, it would also be considered an *in situ* independent variable because the participants are not randomly assigned to the two levels of the independent variable, thus, any implications of causation will be limited because of the *in situ* nature of the design.

A random effects ANOVA design is not often used in psychological research, or the random effects factor is rarely the focus of the experiment. In theory, a **random effects** factor is one where the experimenter is interested in generalizing to other levels of the independent variable, and the levels of the independent variable are randomly chosen from all possible levels of the independent variable. The participants in a repeated measures ANOVA could be considered a random effects factor but they are very rarely the focus of an experiment. The random effects are of concern in advanced topics in ANOVA because random effects require a different error term in the denominator of the F value than the typical within subject's mean square error term.

THE NULL HYPOTHESES IN A TWO FACTOR ANOVA

Three null hypotheses will be tested in a two factor ANOVA: one for each of the main effects and one for the interaction. Each null hypothesis will have a corresponding alternative hypothesis. The final computational analysis will yield three F values, one for each of the two main effects and one for the interaction.

ASSUMPTIONS AND UNEQUAL NUMBERS OF PARTICIPANTS

The assumptions are the same for a factorial ANOVA as for the one factor ANOVA. The dependent variable should come from a population that is normally distributed (normality assumption). The variances of the levels of the independent variables should come from populations whose variances do not differ (homogeneity of variance assumption). Finally, the scores should be independent of one another throughout all levels of the independent variables.

A factorial ANOVA is robust with regards to violations of the first two assumptions but the latter assumption may never be violated. The formulas in this book have been written to handle unequal numbers of participants in each group; however, the first two assumptions become more important when there are unequal numbers of participants in the different conditions. Large numbers of participants in each condition (minimum of 10) and equal numbers of participants in each condition makes the first two assumptions less important.

COMPUTATIONAL EXAMPLE

An experiment was done to compare the effects of anxiety and task difficulty upon performance in a finger maze. There were two levels of anxiety: high and low conditions where the participants were induced to feel a lot of anxiety or very little, and the task to perform was either a simple finger maze or a difficult finger maze. The scores represent the number of errors in performing the finger mazes and, so, better performance is indicated by a low score and poorer performance is indicated by a high score. The maximum number wrong that could be obtained on the task was 15. Seven college students were randomly assigned to serve in each treatment condition for a total of 28 participants.

The data is tabled as follows:

Low anxiety		High anxiety	
Simple task	Hard task	Simple task	Hard task
5	5	3	12
4	10	5	11
7	12	1	15
6	9	4	13
3	11	2	14
5	8	4	12
6	10	5	13

Step 1. The first step is to obtain the Σx and Σx^2 for each group, and the grand total Σx and Σx^2.

$$\Sigma x = \quad 36 \qquad 65 \quad 24 \qquad 90$$

$$\Sigma x^2 = \ 196 \quad 635 \quad 96 \quad 1168$$

Grand Total

$$\Sigma x = 215$$
$$\Sigma x^2 = 2095$$
$$N = 28$$

Step 2. The CM is obtained by squaring the grand total Σx and dividing by N (the total number of observations).

$$CM = \frac{(\text{total } \Sigma x)^2}{N} = \frac{(215)^2}{28} = 1650.893$$

Step 3. The sum of squares total (SST) is obtained by taking the total Σx^2 and subtracting the CM.

$$SST = \text{total } \Sigma x^2 - CM$$

$$SST = 2095 - 1650.893$$

$$SST = 444.107$$

Step 4. First Main Effect: In order to compute the effects of the anxiety factor (low anxiety versus high anxiety), add the sums of the low anxiety conditions together (regardless of the task difficulty), and add the sums of the high anxiety conditions together (regardless of task difficulty).

$$36 + 65 = 101 \quad \text{sum of low anxiety conditions}$$

$$24 + 90 = 114 \quad \text{sum of high anxiety conditions}$$

Next, square these sums and divide by the number of observations upon which these sums were based and add the quotients.

$$\frac{(101)^2}{14} + \frac{(114)^2}{14} = 728.643 = 928.286 = 1656.929$$

Step 5. In order to compute the sum of squares for anxiety (SSA) take the number resulting from Step 4 and subtract the CM.

$$SSA = 1656.929 - CM$$

$$SSA = 1656.929 - 1650.893$$

$$SSA = 6.036$$

Step 6. Second Main Effect: In order to compute the effects of task difficulty (simple task versus hard task), add the sums of the simple task conditions together and add the sums of the hard task conditions together (regardless of anxiety level).

$36 + 24 = 60$ sum of simple task conditions

$65 + 90 = 155$ sum of hard task conditions

Next, square these sums and divide by the number of observations upon which these sums were based and add the quotients.

$$\frac{(60)^2}{14} + \frac{(155)^2}{14} = 257.143 + 1716.071 = 1973.214$$

Step 7. In order to compute the sum of squares for task difficulty (SSD), take the number resulting from Step 6 and subtract the CM.

$SSD = 1973.214 - CM$

$SSD = 1973.214 - 1650.893$

$SSD = 322.321$

Step 8. Interaction: In order to compute the interaction effects between anxiety and task difficulty, take the sum of each of the four conditions or groups, square them, divide them each by the number of observations upon which each was based and add the quotients.

$$\frac{(36)^2}{7} + \frac{(65)^2}{7} + \frac{(24)^2}{7} + \frac{(90)^2}{7} = 185.143 + 603.571 + 82.286 + 1157.143 = 2028.143$$

Step 9. In order to obtain the sum of squares for the interaction effects (SSA × D) take the value obtained from Step 8 and subtract the CM, SSA, and SSD.

$SSA \times D = 2028.143 - CM - SSA - SSD$

$SSA \times D = 2028.143 - 1650.893 - 6.036 - 322.321$

$SSA \times D = 48.893$

Step 10. In order to obtain the error term sum of squares (SSW), take SST and subtract SSA, SSD, and SSA × D.

$SSW = SST - SSA - SSD - SSA \times D$

$SSW = 444.107 - 6.036 - 322.321 - 48.893$

$SSW = 66.857$

Step 11. The *df* are now calculated:

df for SST = total number of scores minus 1

df for SST = 28 – 1

df for SST = 27

df for SSA = total number of anxiety treatment levels minus 1

df for SSA = 2 – 1

df for SSA = 1

df for SSD = total number of task difficulty treatment levels minus 1

df for SSD = 2 – 1

df for SSD = 1

df for SSA × D = the df for SSA × df for SSD

df for SSA × D = 1 × 1

df for SSA × D = 1

df for SSW = df for SST minus df for SSA, SSD, and SSA × D

df for SSW = 27 – 1 – 1 – 1

df for SSW = 24

Step 12. The mean squares are computed by dividing the sum of squares by the appropriate *df*.

$$MSA = \frac{SSA}{df_{SSA}} = \frac{6.036}{1} = 6.036$$

$$MSD = \frac{SSD}{df_{SSD}} = \frac{322.321}{1} = 322.321$$

$$MSA \times D = \frac{SSA \times D}{df_{SSA \times D}} = \frac{48.893}{1} = 48.893$$

$$MSW = \frac{SSW}{df_{SSW}} = \frac{66.857}{24} = 2.786$$

Step 13. The *F* values for testing significance of anxiety, task difficulty and their interaction are obtained by dividing the appropriate mean squares by MSW.

$$\text{F value for anxiety} = \frac{\text{MSA}}{\text{MSW}} = \frac{6.036}{2.786} = 2.17$$

$$\text{F value for task difficulty} = \frac{\text{MSD}}{\text{MSW}} = \frac{322.321}{2.786} = 115.69$$

$$\text{F value for anxiety} \times \text{task difficulty} = \frac{\text{MSA} \times \text{D}}{\text{MSW}} = \frac{48.893}{2.786} = 17.56$$

Step 14. Determine the significance levels of the obtained F values, and table the data as follows:

Source	Sum of squares	df	Mean squares	F
Anxiety	6.036	1	6.036	2.17
Task Difficulty	322.321	1	322.321	115.69*
Anxiety × Task Difficulty	48.893	1	48.89	17.56*
Error	66.857	24	2.79	
Total	444.107	27		

* $p < .01$

Step 15. Make the decision to reject or retain the null hypotheses as follows:

Main Effect: Anxiety

The null hypothesis for the Anxiety factor is:

H_o: $\mu_1 = \mu_2$

H_a: $\mu_1 \neq \mu_2$

where

μ_1 = the mean number of errors for all participants in the low Anxiety condition

μ_2 = the mean number of errors for all participants in the high Anxiety condition

The df_1 for the Anxiety main effect refers to the number of levels of the Anxiety main effect minus 1. This value was obtained in Step 11 (df for SSA). The df_2 or denominator df, refers to the df for SSW, also obtained in Step 11 (df for SSW). Note that in a completely randomized two factor ANOVA, df_2 (denominator df) stays the same for both main effects and the interaction.

Only in this computational example, df_1 and df_2 will be the same for all three F values, that is, $df_1 = 1$ and $df_2 = 24$. The critical value at $p = .05$ with $df_1 = 1$, $df_2 = 24$ is $F = 4.26$. The write-up might look like this:

> The obtained F value for the Anxiety main effect did not exceed the tabled critical value $F = 4.26$ at $p = .05$ with $df_1 = 1$ and $df_2 = 24$. Therefore, H_o is retained, and it is concluded that there was no difference in the mean number of errors for the

participants in the low Anxiety condition versus the mean for those in the high Anxiety condition, $F(1,24) = 2.17$, $p > .05$. In terms of the research question, it appears that there is no effect of anxiety level upon the number of errors.

One final note is that the actual means in the previous write-up are not really necessary or of interest for two reasons: First, there was no significant difference between the two conditions, which makes the finding somewhat less interesting and, second, the interaction was significant so the discussion of the main effects should be minimized.

Main Effect: Task Difficulty
The null hypothesis for the Task Difficulty factor is:

$$H_o: \mu_1 = \mu_2$$

$$H_a: \mu_1 \neq \mu_2$$

where

μ_1 = the mean number of errors for all participants in the simple Task Difficulty condition

μ_2 = the mean number of errors for all participants in the hard Task Difficulty condition

The df_1 for Task Difficulty refers to the number of levels of the Task Difficulty main effect minus 1. This value was obtained in Step 11 (df for SSD). The df_2, or denominator df, refers to the df for SSW, also obtained in Step 11 (df for SSW). The write-up might look like this:

The obtained F value for the Task Difficulty main effect exceeds the critical $F = 4.26$ at $p = .05$ with $df_1 = 1$ and $df_2 = 24$. Therefore, H_o is rejected, and it is concluded that the mean number of errors for all participants in the simple task condition (4.3) was significantly lower than the mean for those in the hard task condition (11.7), $F(1,24) = 115.69$, $p < .01$. In terms of the research question, it appears that the participants make significantly more errors in hard tasks than on simple tasks.

Interaction Effect
The null hypothesis for the interaction between the main effects of Anxiety and Task Difficulty is:

$$H_o: \text{There is no interaction}$$

$$H_a: \text{There is no interaction}$$

where

The df_1 for the interaction is df_1 for the Anxiety main effect *times* the df_1 for the Task Difficulty main effect. This value was also obtained in Step 11 (df for SSA

× D). The df_2, or denominator df, refers to the df for SSW, also obtained in Step 11 (df for SSW). The write-up might look like this:

> The obtained F value for the interaction effect exceeds the critical $F = 4.26$ at $p = .05$ with $df_1 = 1$ and $df_2 = 24$. Therefore, H_o is rejected, and it is concluded that at least one mean is significantly different from one other mean, $F (1,24) = 17.56$, $p < .01$. In order to determine the pattern of mean differences, a multiple comparison test must be performed. In terms of the research question, there is a significant interaction between anxiety level and task difficulty upon the number of errors in performing the mazes.

WARNING: LIMIT YOUR MAIN EFFECT CONCLUSIONS WHEN THE INTERACTION IS SIGNIFICANT

If the interaction is significant in a two factor ANOVA, there was some unique effect of the two variables together upon the dependent variable that *could not have been predicted* from either of the two main effect conclusions independently. Thus, if the interaction is significant, limit your conclusions severely. In fact, some statisticians recommend making *no conclusions at all* about the main effects if the interaction is significant.

MULTIPLE COMPARISON TESTS

If the anxiety main effect had been significant, no multiple comparison test would have been necessary. Since there were only two means in the alternative hypothesis, we would know immediately which mean was significantly higher or lower than the other. A multiple comparison test would only be performed if the *F* value was significant, *and* there were three or more means in the alternative hypothesis. A significant interaction always requires a multiple comparison test.

Most multiple comparison tests may be performed at the $p < .05$ or $p < .01$ level, regardless of whether the main effects or interaction met the $p < .01$ level or not. Your guiding principle in conducting multiple comparison tests should be which p level explains your data the best.

WRITING UP THE RESULTS JOURNAL STYLE

Journal article versions of a two factor ANOVA may actually be rather brief. For our previous example, it might look like this:

> The main effect of Anxiety was not significant ($F (1,24) = 2.17$, $p > .05$) while the main effect of Task Difficulty ($F (1,24) = 115.69$, $p < .01$) and the interaction ($F (1,24) = 17.56$, $p < .01$) were significant. Tukey's test of the interaction revealed that . . .

Once again, the most important feature of a two factor ANOVA is the interaction. A significant interaction means that the results could not have been predicted by interpreting the main effects alone. In fact, if a significant interaction is present, the interpretation of the two main effects could be *misleading*. If there is no significant interaction then the main effects can be interpreted without any problem in interpretation.

LANGUAGE TO AVOID

Statisticians are surprisingly touchy about certain words and phrases. Some people use the phrase 'the null hypothesis was accepted' which seems to send nearly every statistician into a tizzy (a tizzy looks like a huff, only less serious). 'The null hypothesis was accepted' has the implication that the null hypothesis is true or that we are endorsing its veracity when this is not really the case. We may not reject the null hypothesis for a variety of reasons *NOT* related to its truth or falsity. For example, the experiment may simply lack power (too few participants), the independent variable may have too weak an effect, or the dependent variable may be inappropriate or insensitive. Some statisticians do not even like the phrase 'retain or reject the null hypothesis.' They feel that even the word 'retain' has the connotation of endorsing the truth of the null hypothesis. These statisticians prefer the phrase 'reject or fail to reject the null hypothesis.'

Another phrase that annoys some statisticians is 'highly significant.' They argue that the null hypothesis is either rejected or not. Since the conventional level of significance is $p = .05$, the null hypothesis is rejected if it exceeds the critical value at this p level. They argue that we do not reject the null hypothesis at $p < .01$ or $p < .001$. It has already been rejected at $p = .05$. So, for example, when we report a finding significant at $p < .001$, it is not 'highly significant.' It is simply significant (the null hypothesis has been rejected at $p = .05$) and the probability of the Type I Error is very low (*i.e.*, less than one chance in 1000 for $p < .001$). The phrase 'highly significant' may also confuse the difference between the use of the word 'significant' as a value judgment (whether something is worthwhile or valuable) or as a statistical decision (findings are significant if the null hypothesis has been rejected). The phrase 'highly significant' has the connotation of a value judgment combined with a statistical decision which would be inappropriate. It would be best if we learn to say 'a significant finding with a low probability of the Type I Error.'

EXPLORING THE POSSIBLE OUTCOMES IN A TWO FACTOR ANOVA

Since there are two main effects and an interaction effect in a two factor ANOVA, and there are only two possible hypothesis testing outcomes (significant or not), then there are $2 \times 2 \times 2 = 8$ total possible outcomes. See Figure 12.1 for a summary of these outcomes. Note that the interaction is significant in every case where the lines for the main effects are not parallel.

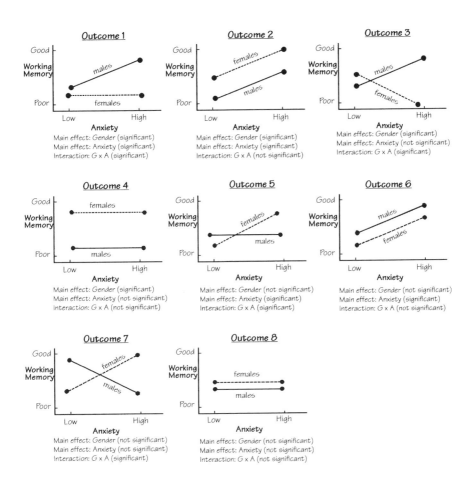

Figure 12.1 Eight Possible Main Effects and Interaction Outcomes of a Two Factor ANOVA

Note: In this hypothetical example of a two factor ANOVA, Gender (males vs. females) and Anxiety level (low or high) were the main effects. The dependent variable was working memory (short-term memory capacity).

DETERMINING EFFECT SIZE IN A TWO FACTOR ANOVA

Since a significant or nonsignificant F value still does not reveal the strength of the independent variable, we can determine the effect size by omega-squared (ω^2) just as in the one factor ANOVA but it requires three separate analyses, one for each F value. For a two factor ANOVA, the formulas are as follows:

For the first main effect (labeled Factor A):

$$\omega_A^2 = \frac{SSA - (k-1)MSW}{SST + MSW}$$

where

    SSA   = the sum of the squares for the first main effect or main effect A

    k   = the number of levels of the first main effect

For the second main effect (labeled Factor B):

$$\omega_B^2 = \frac{SSB - (k-1)MSW}{SST + MSW}$$

where

    SSB   = the sum of squares for the second main effect or main effect B

    k   = the number of levels of the second main effect

For the interaction (labeled A × B):

$$\omega_{A \times B}^2 = \frac{SSA \times B - (df \text{ for interaction})MSW}{SST + MSW}$$

where

    SSA × B   = the sum of squares for the interaction

    $df_{\text{for interaction}}$ = df for the first main effect times df for the second main effect

In most cases, we will only be concerned with effect size determination when an F value is significant. However, when the power of an ANOVA is low (perhaps because of too few participants), and the obtained F value did not reach significance but was close (called a trend), you may be interested in determining effect size in order to see if the independent variable did have an impact on the dependent variable despite the lack of significance.

In the previous computational example, for the Anxiety main effect:

$$\omega^2_A = \frac{SSA - (k-1)MSW}{SST + MSW}$$

where

SSA   = the sum of squares for the Anxiety factor

k   = the number of levels of the Anxiety factor

$$\omega^2_A = \frac{6.036 - (2-1)2.79}{444.107 + 2.79} = 0.007$$

In the previous computational example, for the Task Difficulty main effect:

$$\omega^2_B = \frac{322.321 - (2-1)2.79}{444.107 + 2.79} = 0.715$$

In the previous computational example, for the interaction:

$$\omega^2_{A \times B} = \frac{48.893 - (1)2.79}{444.107 + 2.79} = 0.103$$

According to omega-squared interpretation guidelines,

$\omega^2 > .15$ = large effect

$\omega^2 > .06$ = medium effect

$\omega^2 > .01$ = small effect

Thus, we can see that for the Anxiety main effect, the effect size determination supported the nonsignificant findings because the $\omega^2 = .007$ indicates a less than small effect size. The main effect for Task Difficulty has a whoppingly large effect size with $\omega^2 = .715$. Remember that ω^2 may also be interpreted as the percentage of variance in the dependent variable accounted for by the independent variable. ω^2 is a measure of how much variation is accounted for by the treatment and, by default, how much of the total variation in scores is accounted for by random factors. For the Task Difficulty main effect, 72% of the total variance in the scores can be accounted for by the task difficulty and 28% of the total variance is accounted for by unknown or random factors. For the interaction, the $\omega^2 = .103$ is indicative of a medium to large effect size and, thus, the interaction between the two main effects accounts for about 10% of the variance in the dependent variable.

Factorial Design – a design in ANOVA where two or more main effects and the interaction between the main effects can be evaluated.

Completely Randomized Factorial Design – a design in ANOVA where the participants are randomly assigned to the conditions. *In situ* ANOVA designs may actually be more common, and the former design allows for the implication of causation while the latter does not.

Main Effect – another name for the independent variable in ANOVA designs. Each main effect must have two or more levels.

Interaction Effect – refers to the unique confluence of two main effects or independent variables upon each other in factorial designs. The interaction effect may reveal outcomes that could not have been predicted had the experiment consisted of two separate one factor ANOVAs of the same independent variables.

Simple Effects – refers to the situation where the interaction is significant and the experimenter is interested in all levels of one independent variable under just one level of the other independent variable.

1. A gerontologist wishes to determine whether older adults differ from younger adults in a complex reaction time experiment involving three levels of a minor tranquilizer (Valium). Fifteen older adults (mean age 65 years, range = 60 to 75 years) were compared to fifteen young adults (mean age 21 years, range = 18 to 23) on a computer task measuring reaction time. The reaction times varied on the task from 2 seconds (a very quick reaction time or a good score) to over 10 seconds (a very slow reaction time or bad score). The participants were randomly assigned to receive either 0 milligrams (placebo), 5 mgs, or 10 mgs, and one-half hour later they were tested for their reaction times in total seconds (rounded to whole numbers).

Young			Old		
0	5	10	0	5	10
7	7	13	5	4	18
4	8	10	7	3	16
5	7	10	5	2	16
5	6	9	6	6	20
4	7	10	4	4	14

2. A psychologist wishes to determine whether short-term memory scores are affected by the type of medication in hyperactive children. Sixty hyperactive children (30 boys and 30 girls) were randomly assigned to receive either Cylert or Ritalin (two kinds of amphetamines which are CNS stimulants). Their short-term memory was tested on a 20 point scale (where 0 = no memory and 20 = perfect memory). Is there any evidence that the drugs help or hurt memory? Discuss your findings!

Boys			Girls		
Cylert	Placebo	Ritalin	Cylert	Placebo	Ritalin
12	5	3	5	4	12
10	6	2	6	6	9
13	7	5	2	7	14
10	6	6	4	6	13
9	4	1	3	5	9
12	5	4	2	4	12
13	6	2	4	5	8
10	7	6	6	6	15
11	5	1	2	7	12
12	7	5	4	7	11

3. The element mercury causes massive brain dysfunction, including hallucinations and delusions. Also, a 'silver filling' from your dentist is actually 50–66% mercury and only 35% silver. The American Dentistry Association (ADA) says that 'silver amalgams' are safe because the mercury is not released. However, mercury can be measured in your hair or your urine because the mercury is apparently released by chewing. The ADA says this amount is not bothersome. A psychologist wishes to determine whether mercury fillings affect a child's brain functioning. An MBD (minimal brain dysfunction) parental survey-test is created which has twenty questions about MBD symptoms like 'My child has learning problems,' 'My child is hyperactive,' etc. The overall score on the MBD test ranges from 0 (no evidence of brain dysfunction) to 20 (severe brain dysfunction). The survey is given to parents as they visit a dentist with their child. The number of 'silver' fillings is noted, as is the child's gender. For analysis purposes, the children are divided up into whether they have no fillings, one to three fillings, or more than five fillings, and their MBD score is noted. Is gender a factor on the MBD scale? Does the number of fillings have an effect on the MBD scale? Is there an interaction? Discuss and criticize your findings!

No fillings		One to three fillings		Five or more fillings	
Male	Female	Male	Female	Male	Female
7	4	11	6	17	13
4	6	12	9	20	15
6	7	10	9	15	11
7	8	9	9	17	12
8	4	13	8	16	10
5	3	14	5	14	12
9	7	15	7	18	14
5	9	10	5	20	14

4. For Problems 1, 2, and 3, identify whether the independent variables in each design are fixed or random factors, and whether the factors are *in situ* or not.

Problem 1.

Age:

Drug:

Problem 2.

Gender:

Drug:

Problem 3.

Element:

Gender:

13 Factorial Analysis of Variance: Additional Designs

THE SPLIT-PLOT DESIGN

The previous chapter introduced the completely randomized factorial ANOVA. The present chapter will present two additional two factor ANOVA designs: the **split-plot ANOVA** and the **repeated measures ANOVA**. Chapter 12 introduced the two factor completely randomized ANOVA where both of the independent variables were considered to be **between-groups** variables, *i.e.*, different participants were assigned to each level of each factor. The split-plot design is a combination of a completely randomized design and a repeated measures design. It requires that there is one between-groups variable (different participants in each level of the independent variable) and one **within-subjects** variable (the same participant serves in each level of the other independent variable). Frequently, the within-subjects variable consists of taking measurements of participants across time or over successive trials. Interestingly, the split-plot design takes its name from the historical roots of statistics in agriculture. Remember that Gossett and Fisher both worked on statistics related to the growing of crops (in Gossett's case, wheat, barley, and malt used in the brewing of beer). The notion of splitting a plot of land and trying different growing techniques (*e.g.*, fertilizers, watering cycles, etc.) on each plot led to the statistical designs being labeled 'split-plot.'

A two factor repeated measures ANOVA has two factors with two or more levels of each factor, and the same group of participants serves in every level of both factors. The design has the advantage of using fewer participants than the two factor ANOVA without a loss of power. However, there are many experimental designs where the repeated measures design is unfeasible. The two factor repeated measures design will be discussed in greater detail later in this chapter.

OVERVIEW OF THE SPLIT-PLOT ANOVA

The split-plot ANOVA has three F values for the two main effects and the interaction just as in the two factor completely randomized ANOVA. The difference between the two designs lies in the error term. In the two factor completely randomized ANOVA, there was only one error term for both main effects and the interaction. The denominator in the F ratio was the same for all three effects. In the two factor split-plot ANOVA, we will derive two different error terms: one error term for the between-subjects variable and the other error term for the within-subjects factor and the interaction. Otherwise, the assumptions are the same for both designs as are the null and alternative hypotheses, significance testing, and the interpretation.

COMPUTATIONAL EXAMPLE

A social facilitation experiment was conducted where there were two levels of a social facilitation factor: a participant working alone in a room and a participant working with another participant (a confederate) present in a room. The task required the participant to judge different weights and order them according to magnitude. The other person present in the room only watched and did not speak or participate. Each participant served for three trials and three different ranges of weights were randomized across participants, so that each participant received a different weight range on each trial. Six participants were randomly assigned to each of the two social facilitation conditions for a total of 12 participants. Time and number of errors were combined in a special manner for one overall score per trial for each participant, and lower scores are better scores. The data is tabled as follows:

Participants alone				Participants observed			
	Trial				Trial		
	1	2	3		1	2	3
P_1	9	7	7	P_7	6	7	5
P_2	8	9	7	P_8	8	6	5
P_3	10	6	8	P_9	5	5	4
P_4	7	8	8	P_{10}	6	6	5
P_5	9	8	8	P_{11}	7	5	4
P_6	8	9	8	P_{12}	6	6	4

Step 1. Obtain Σx and Σx^2 for each treatment combination and the overall Σx and Σx^2.

	Alone			Observed		
Trials	1	2	3	1	2	3
Σx	51	47	46	38	35	27
Σx^2	439	375	354	246	207	123

Overall $\Sigma x = 244$
Overall $\Sigma x^2 = 1744$

Step 2. The CM is obtained by squaring the overall Σx and dividing by the total number of observations.

$$CM = \frac{(\text{overall } \Sigma x)^2}{N} = \frac{(244)^2}{36} = 1653.78$$

Step 3. The sum of squares total (SST) is obtained by taking the overall Σx^2 and subtracting the CM.

$SST = \text{overall } \Sigma x^2 - CM$

$SST = 1744 - 1653.78$

$SST = 90.22$

Step 4. In order to calculate the between-subjects sum of squares (SSB), add the three trial scores together for each participant.

Alone		Observed	
P_1	23	P_7	18
P_2	24	P_8	19
P_3	24	P_9	14
P_4	23	P_{10}	17
P_5	25	P_{11}	16
P_6	25	P_{12}	16

Next, square each of these sums and add them together.

$23^2 + 24^2 + 24^2 + 23^2 + 25^2 + 25^2 + 18^2 + 19^2 + 14^2 + 17^2 + 16^2 + 16^2$
$= 5142$

Divide this obtained value by the number of trials summed across and subtract the CM to obtain SSB.

$$SSB = \frac{5142}{3} - CM$$

$SSB = 1714 - 1653.78$

$SSB = 60.22$

Step 5. In order to calculate the sum of squares for the social facilitation factor (SSF), obtain Σx for each level of the social facilitation factor, square each and divide each by the number of scores within each group and add the quotients.

sum of participants alone $= 51 + 47 + 46 = 144$

sum of participants with observer $= 38 + 35 + 27 = 100$

$$\frac{(144)^2}{18} + \frac{(100)^2}{18} = 1707.56$$

From this value, subtract the CM in order to obtain SSF.

SSF = 1707.56 – CM

SSF = 1707.56 – 1653.78

SSF = 53.78

Step 6. In order to compute the sum of squares error term (SSBW) for the between-subjects factor, subtract SSF from SSB.

SSBW = SSB – SSF

SSBW = 60.22 – 53.78

SSBW = 6.44

Step 7. In order to compute the within-subjects sum of squares (SSW), subtract SSB from SST.

SSW = SST – SSB

SSW = 90.22 – 60.22

SSW = 30.00

Step 8. In order to compute the sum of squares for trials (SSTR) add the scores of trial 1 across both social facilitation conditions, add the scores of trial 2 and the scores of trial 3.

Trials $1 + 2 + 3$

Σx^2 $= 89 + 82 + 73$

Next, square each of these sums and divide these values by the number of scores upon which each sum was based and add the quotients.

$$\frac{89^2}{12} + \frac{82^2}{12} + \frac{73^2}{12} = 1664.50$$

From this value, subtract the CM in order to obtain SSTR.

SSTR = 1664.50 – CM

SSTR = 1664.50 – 1653.78

SSTR = 10.72

Step 9. In order to compute the interaction sum of squares between trials and social facilitation (SSF × TR), obtain the sums for each combination of the trials and social facilitation factors.

	Alone			Observed		
Trial	1	2	3	1	2	3
Σx	51	47	46	38	35	27

Next, square each of these values and divide each by the number of scores upon which each sum was based and add the quotients.

$$\frac{51^2}{6} + \frac{47^2}{6} + \frac{46^2}{6} + \frac{38^2}{6} + \frac{35^2}{6} + \frac{27^2}{6} = 1720.67$$

From this value, subtract the CM, SSF, and SSTR in order to obtain SSF × TR.

SSF × TR = 1720.67 – CM – SSF – SSTR

SSF × TR = 1720.67 – 1653.78 – 53.78 – 10.72

SSF × TR = 2.39

Step 10. In order to compute the within-subjects sum of squares error term (SSB × W) for the within-subjects main effect and the interaction, take SSW and subtract SSTR and SSF × TR.

SSB × W = SSW – SSTR – SSF × TR

SSB × W = 30.00 – 10.72 – 2.39

SSB × W = 16.89

Step 11. The *df* are now calculated.

df for SST = total number of scores – 1

= 36 – 1

df for SST = 35

df for SSB = total number of participants – 1

= 12 – 1

df for SSB = 11

df for SSF = total number of levels of the social facilitation factor – 1

= 2 – 1

df for SSF = 1

df for SSBW = *df* for SSB – *df* for SSF

$$= 11 - 1$$

df for SSBW $\quad = 10$

df for SSW $\quad = df$ for SST $- df$ for SSB

$$= 35 - 11$$

df for SSW $\quad = 24$

df for SSTR $\quad =$ the number of trials (for each participant) -1

$$= 3 - 1$$

df for SSTR $\quad = 2$

df for SSF $\times$ TR $= df$ for SSTR $\times df$ for SSF

$$= 2 \times 1$$

df for SSF $\times$ TR $= 2$

df for SSB $\times$ W $= df$ for SSW $- df$ for SSTR $- df$ SSF $\times$ TR

$$= 24 - 2 - 2$$

df for SSB $\times$ W $= 20$

Step 12. The mean squares (MS) are computed by dividing the appropriate sum of squares by the appropriate df for each sum of squares.

$$\text{MSF} = \frac{\text{SSF}}{df_{\text{for SSF}}} = \frac{53.78}{1} = 53.78$$

$$\text{MSBW} = \frac{\text{SSBW}}{df_{\text{for SSBW}}} = \frac{6.44}{10} = 0.644$$

$$\text{MSTR} = \frac{\text{SSTR}}{df_{\text{for SSTR}}} = \frac{10.72}{2} = 5.36$$

$$\text{MSF} \times \text{T} = \frac{\text{SSF} \times \text{TR}}{df_{\text{for SSF} \times \text{TR}}} = \frac{2.39}{2} = 1.20$$

$$\text{MSB} \times \text{W} = \frac{\text{SSB} \times \text{W}}{df_{\text{for SSB} \times \text{W}}} = \frac{16.89}{20} = 0.85$$

Step 13. The appropriate F values are computed by the following formulae.

Main Effect: Social Facilitation

$$F = \frac{\text{MSF}}{\text{MSBW}} = \frac{53.78}{0.644} = 83.51$$

Main Effect: Trials

$$F = \frac{MSTR}{MSB \times W} = \frac{5.36}{0.85} = 6.31$$

Interaction: Social Facilitation × Trials

$$F = \frac{MSF \times TR}{MSB \times W} = \frac{1.20}{0.85} = 1.42$$

Step 14. In order to check these obtained F values for significance, refer to the Appendix of the F distribution. To enter the table, the df_1 is the df for the appropriate numerator sum of squares and df_2 is the df for the error term used to test the respective sum of squares.

Thus,

social facilitation factor	$df_1 = 1,$	$df_2 = 10$
trials factor	$df_1 = 1,$	$df_2 = 20$
social facilitation and trials	$df_1 = 1,$	$df_2 = 20$

Step 15. After checking for significance, table the data as follows:

Source	SS	df	MS	F	p
Social Facilitation	53.78	1	53.78	83.51	< .001
Error (SSBW)	6.44	10	.64		
Trials (TR)	10.72	2	5.36	6.31	< .01
TR × F	2.39	2	1.20	1.42	> .05
Error (SSB × W)	16.89	20	.85		
Total	90.22	35			

Step 16. As in all two factor ANOVA designs, there are three null and alternative hypotheses.

Main Effect: Social Facilitation

In this example, the null and alternative hypotheses for the social facilitation factor are as follows:

$H_o: \mu_{alone} = \mu_{observed}$

$H_o: \mu_{alone} \neq \mu_{observed}$

Since the F value is significant for the Social Facilitation factor, H_o is rejected. There are only two means associated with the rejected H_o; no multiple comparison test is necessary. An inspection of the means reveals that the mean number of errors (5.6) of the participants who were observed is significantly lower than the mean errors (8.0) of the unobserved participants $F(1,10) = 83.5, p < .001$.

Main Effect: Trials

The Trials factor was also significant. The H_o and H_a are:

H_o: $\mu_{trial\ 1}$ $\mu_{trial\ 2} = \mu_{trial\ 3}$

H_a: H_o is not true

Since the Trials factor null hypothesis was rejected and because it contains more than two means, a multiple comparison test is necessary to determine which of the means are significantly different from each other. From the ANOVA it can only be concluded that at least one trial mean is significantly different from at least one other trial mean. The pattern of mean differences can only be revealed with a multiple comparison test. The F may be reported as:

$F(2,20) = 6.31, p < .01$

Interaction: Social Facilitation x Trials

The interaction between Social Facilitation and the Trials factor was not significant. The null and alternative hypotheses are:

H_o: There is no interaction

H_a: There is no interaction

Since the obtained F for the interaction did not exceed the critical value of F at $p = .05$, H_o is retained, and no multiple comparison test is necessary nor appropriate. It can be concluded that there is no significant interaction between the Social Facilitation main effect and the Trials main effect on the number of errors. The F can be reported as:

$F(2,20) = 1.42, p > .05$

TWO FACTOR ANOVA: REPEATED MEASURES ON BOTH FACTORS DESIGN

The two factor repeated measures design is appropriate for the analysis of two factors with two or more levels of each factor and where a single group of participants serves in every level of every factor. One advantage of this design is that it uses fewer participants than in a two factor completely randomized ANOVA. However, there are many circumstances where this design cannot be used, e.g., where the two conditions of one factor are inherently fixed (such as *in situ* designs) like in the factor of gender. If the participants must remain naive under all conditions, then different participants and a between-subjects design will be necessary. Also, it is important to note whether one factor may influence the outcome of the other factor. In this case, the assumption of independence between factors will be violated, and the two factor repeated measures ANOVA is completely inappropriate. The assumptions for the repeated measures factorial ANOVA are the same as for the split-plot ANOVA.

OVERVIEW OF THE REPEATED MEASURES ANOVA

The repeated measures ANOVA is similar to the completely randomized and split-plot ANOVA designs. There will be three F values for the two main effects and the interaction. The assumptions also remain the same, but the assumption of independence between the factors is essential. While the assumptions of normality and homogeneity of variance may be violated to some extent because of the robustness of ANOVA, the assumption that scores under one factor are independent of the other factor is sacrosanct.

Computationally, one interesting feature of the two factor repeated measures ANOVA is that each main effect and the interactions have their own unique error term. Thus, the denominators of the three F values are all different.

COMPUTATIONAL EXAMPLE

An experimenter was interested to see whether chimpanzees varied in learning ability from morning to evening. A maze learning task was employed and the number of errors (false turns) was recorded for three trials. Different mazes were used in the morning and evening conditions and prior testing indicated that they were of equal difficulty. The data is tabled as follows:

	Morning				**Evening**		
Trial	1	2	3	Trial	1	2	3
Chimp 1	6	5	6		3	7	5
Chimp 2	8	4	4		5	5	6
Chimp 3	6	6	7		4	6	8
Chimp 4	7	8	7		4	4	1
Chimp 5	7	6	7		4	6	6
Chimp 6	6	5	6		4	6	6
Chimp 7	8	5	9		5	7	4
Chimp 8	7	6	6		4	6	5
Chimp 9	9	5	8		5	3	5
Chimp 10	6	7	6		4	5	3

Step 1. Obtain Σx and Σx^2 for each treatment combination and the overall Σx and Σx^2 for each participant's scores.

	Morning				Evening				
Trial	1	2	3	Row Σx	1	2	3	Row Σx	Total Row Σx
C_1	6	5	6	17	3	7	5	15	32
C_2	8	4	4	16	5	5	6	16	32
C_3	6	6	7	19	4	6	8	18	37
C_4	7	8	7	22	4	4	1	19	31
C_5	7	6	7	20	4	6	6	16	36
C_6	6	5	6	17	4	6	6	16	33
C_7	8	5	9	22	5	7	4	16	38
C_8	7	6	6	19	4	6	5	15	34
C_9	9	5	8	22	5	3	5	13	35
C_{10}	6	7	6	9	4	5	3	12	31
Σx	70	57	66		42	55	49		
Σx^2	500	337	452		180	317	273		

Total $\Sigma x = 339$
Total $\Sigma x^2 = 2059$

Step 2. To obtain the CM, square the overall Σx and divide by the total number of observations.

$$\mathrm{CM} = \frac{(\text{overall } \Sigma x)^2}{N} = \frac{(339)^2}{60} = 1915.35$$

Step 3. To obtain the sum of squares total (SST), subtract the CM from the overall Σx^2.

$$\mathrm{SST} = \text{overall } \Sigma x^2 - \mathrm{CM}$$

$$\mathrm{SST} = 2059 - 1915.35$$

$$\mathrm{SST} = 143.65$$

Step 4. To obtain the sum of squares for subjects (SSS), obtain the sum of each participant's scores, square them, and add these squared values together. Divide this summed value by the number of scores summed across (how many scores went into an individual's sum) and subtract the CM.

$$32^2 + 32^2 + 37^2 + 31^2 + 36^2 + 33^2 + 38^2 + 34^2 + 35^2 + 31^2 = 11{,}549$$

$$\mathrm{SSS} = \frac{11{,}549}{6} = 1924.4 - \mathrm{CM}$$

$$\mathrm{SSS} = 1924.4 - 1915.35$$

$$\mathrm{SSS} = 9.48$$

Step 5. To calculate sum of squares for the circadian (morning versus evening) factor (SSC), sum the scores for all the morning trials and then sum all the scores for the evening trials.

$70 + 57 + 66 = 193$ sum for morning trials

$42 + 55 + 49 = 146$ sum for evening trials

Square these sums, add the resulting values together, and divide by the number of scores summed across to obtain an individual sum.

$$(193)^2 + (146)^2 = \frac{58.565}{30} = 1952.17$$

To obtain SSC subtract the CM from this resulting value.

$SSC = 1952.17 - CM$

$SSC = 1952.17 - 1915.35$

$SSC = 36.82$

Step 6. To obtain the sum of squares for the trials factor (SSTR), sum all the scores for Trial 1, sum all the scores for Trial 2, and sum the scores for Trial 3.

$70 + 42 = 112$ sum for Trial 1

$57 + 55 = 112$ sum for Trial 2

$66 + 49 = 115$ sum for Trial 3

Now square each of these sums and add them together, and divide this resulting value by the number of scores summed across for an individual trial.

$$(112)^2 + (112)^2 + (1.15)^2 = 38,313$$

$$\frac{38,313}{20} = 1915.65$$

Now, subtract the CM from this resulting value to obtain SSTR.

$SSTR = 1915.65 - CM$

$SSTR = 1915.65 - 1915.35$

$SSTR = .30$

Step 7. In order to obtain the interaction sum of squares (SSC × TR) take the Σx for each treatment combination, square each of these values, divide each by the number of scores that went into each Σx and add the quotients.

$$\frac{(70)^2}{10} + \frac{(57)^2}{10} + \frac{(66)^2}{10} + \frac{(42)^2}{10} + \frac{(55)^2}{10} + \frac{(49)^2}{10} = 1969.5$$

Step 8. In order to obtain $SSC \times TR$, subtract from the value obtained in Step 7, the CM, SSC, and SSTR.

$$SSC \times TR = 1969.5 - CM - SSC - SSTR$$

$$SSC \times TR = 1969.5 - 1915.35 - 36.82 - .30$$

$$SSC \times TR = 17.03$$

Step 9. In order to calculate the error term for the circadian factor (SSCW), obtain Σx for each participant for each level of the circadian factor (ignoring the trials factor), square these sums, add them together and divide by the number of scores added to get each of the individual sums.

	Circadian Factor	
	Morning	Evening
C_1	6 + 5 + 6	3 + 7 + 5
C_2	16	16
C_3	19	18
C_4	22	9
C_5	20	16
C_6	17	16
C_7	22	16
C_8	19	15
C_9	22	13
C_{10}	19	12

Thus,

$$17^2 + 16^2 + 19^2 + 22^2 + 20^2 + 17^2 + 22^2 + 19^2 + 18^2 + 16^2 + 18^2 + 9^2 + 16^2 + 16^2 + 16^2 + 15^2 + 13^2 + 12^2$$

$$\frac{5961}{3} = 1987.00$$

Step 10. To obtain the within-subjects error term (SSCW), take the value obtained in Step 9, and subtract the CM, SSS, and SSC.

$$SSCW = 1987 - CM - SSS - SSC$$

$$SSCW = 1987 - 1915.35 - 9.48 - 36.82$$

$$SSCW = 25.35$$

Step 11. In order to calculate the error term for the trials factor (SSTRW), first obtain Σx for each participant for each level of the trials factor (ignoring the circadian factor), square these sums, add them together, and divide by the number of scores added to get each of the individual sums.

	Trials		
	1	2	3
C_1	6 + 3	5 + 7	6 + 5
C_2	13	9	10
C_3	10	12	15
C_4	11	12	8
C_5	11	12	13
C_6	10	11	12
C_7	13	12	13
C_8	11	12	11
C_9	14	8	13
C_{10}	10	12	9

Thus,

$$9^2 + 12^2 + 11^2 + 13^2 + 9^2 + 10^2 + 12^2 + 15^2 + 11^2 + 12^2 + 8^2 + 11^2 + 12^2 + 13^2 + 11^2 + 12^2 + 11^2 + 14^2 + 8^2 + 13^2 + 10^2 + 12^2 + 9^2$$

$$\frac{3915}{2} = 1957.50$$

Step 12. In order to obtain SSTRW, take the value obtained in Step 11, and subtract the CM, SSS, and SSTR.

$$SSTRW = 1957.5 - CM - SSS - SSTR$$

$$SSTRW = 1957.5 - 1915.35 - 9.48 - .30$$

$$SSTRW = 32.37$$

Step 13. In order to obtain the error term for the interaction (SSC × TRW), take SST and subtract from it, SSS, SSC, SSTR, SSC × TR, SSTRW, and SSCW.

$$SSC \times TRW = SST - SSS - SSC - SSTR - SSC \times TR - SSTRW - SSCW$$

$$SSC \times TRW = 143.65 - 9.48 - 36.82 - .30 - 17.03 - 25.35 - 32.37$$

$$SSC \times TRW = 22.30$$

Step 14. The *df* are now calculated.

df for SST = total number of scores − 1

$$= 60 - 1$$

df for SST = 59

df for SSS = the number of participants − 1

$= 10 - 1$

df for SSS $= 9$

df for SSC $=$ number of levels of the circadian factor $- 1$

$= 2 - 1$

df for SSC $= 1$

df for SSTR $=$ number of levels of trials factor $- 1$

$= 3 - 1$

df for SSTR $= 2$

df for SSC $\times$ TR $= df$ for SSC times df for SSTR

$= 1 \times 2$

df for SSC $\times$ TR $= 2$

df for SSCW $= df$ for SSS times df for SSC

$= 9 \times 1$

df for SSCW $= 9$

df for SSTRW $= df$ for SSS times df for SSTR

$= 9 \times 2$

df for SSTRW $= 18$

df for SSC $\times$ TRW $= df$ for SSS times df for SSC times df for SSTR

$= 9 \times 1 \times 2$

df for SSC $\times$ TRW $= 18$

Step 15. The mean squares MS are now calculated by dividing the sum of squares by its appropriate df (see Step 14).

$$\text{MSC} = \frac{\text{SSC}}{df_{SSC}} = \frac{36.82}{1} = 36.82$$

$$\text{MSTR} = \frac{\text{SSTR}}{df_{SSTR}} = \frac{0.30}{2} = 0.15$$

$$\text{MSC} \times \text{TR} = \frac{\text{SSC} \times \text{TR}}{df_{SSC \times TR}} = \frac{17.03}{2} = 8.52$$

$$\text{MSCW} = \frac{\text{SSCW}}{df_{SSCW}} = \frac{25.35}{9} = 2.82$$

$$\text{MSTR} = \frac{\text{SSTRW}}{df_{SSTRW}} = \frac{32.37}{18} = 1.80$$

$$\text{MSC} \times \text{TRW} = \frac{\text{SSC} \times \text{TRW}}{df_{SSC \times TRW}} = \frac{22.30}{18} = 1.24$$

Step 16. The appropriate F values are now calculated by the following formulae:

Main Effect: Circadian Factor

$$F = \frac{\text{MSC}}{\text{MSCW}} = \frac{36.82}{2.82} = 13.06$$

Main Effect: Trials Factor

$$F = \frac{\text{MSTR}}{\text{MSTRW}} = \frac{0.15}{1.85} = 0.08$$

Interaction: Circadian $\times$ Trials

$$F = \frac{\text{MSC} \times \text{TR}}{\text{MSC} \times \text{TRW}} = \frac{8.52}{1.24} = 6.87$$

Step 17. In order to check these obtained F values for significance, refer to F distribution in the appendix. To enter the table, df_1 is the df associated with the numerator or sum of squares in the obtained F value, and df_2 is the df for the error term or the denominator in the obtained F value.

Thus,

circadian factor	$df_1 = 1, df_2 = 9$
trials factor	$df_1 = 2, df_2 = 18$
interaction factor	$df_1 = 2, df_2 = 18$

Step 18. After checking for significance, table the data as follows:

Source	SS	df	MS	F	p
Circadian (SSC)	36.82	1	36.82	13.06	< .01
Error (SSCW)	25.35	9	2.82		
Trials (SSTR)	.30	2	.15	.08	> .05
Error (SSTRW)	32.37	18	1.80		
C × TR (SSC × TR)	17.03	2	8.52	6.80	< .01
Error (SSC × TRW)	22.30	18	1.24		
Subjects	9.48	9	1.05		
Total	143.65	59			

Step 19. As in all two factor ANOVA designs, there are three null and alternative hypotheses. In this example, there is an H_o and H_a for the Circadian factor:

$$H_o: \mu_{morning} = \mu_{evening}$$

$$H_a: \mu_{morning} \neq \mu_{evening}$$

Since the F value is significant for the Circadian Factor, H_o is rejected. There are only two means associated with the rejected H_o, so no multiple comparison test is necessary. An inspection of the means reveals that the mean number of errors by the chimps (6.4) in the morning was significantly lower than their mean errors (11.3) in the evening. In a report, the F may be reported as:

$$F(1,9) = 13.06, p < .01$$

The Trials Factor was not significant. The H_o and H_a are:

$$H_o: \mu_{Trial\ 1} = \mu_{Trial\ 2} = \mu_{Trial\ 2}$$

$$H_a: H_o \text{ is not true}$$

Since the Trials Factor H_o was retained, no multiple comparison test is necessary. It may be concluded that the means for the three trials do not differ significantly from one another. The F may be reported as:

$$F(2,18) < 1.00, p > .05$$

The interaction between Circadian Factor and the Trials Factor was significant. The null and alternative hypotheses are:

H_o: There is no interaction.

H_a: There is no interaction.

Since the obtained F for the interaction exceeded the critical value of F at $p < .05$, H_o is rejected, and it may be concluded that at least one of these means is significantly different from one other. To determine the pattern of differences between the six means, a multiple comparison test is necessary. The F is reported as:

$$F(2,18) = 8.52, \quad p < .01$$

Split-Plot ANOVA – refers to a two factor ANOVA where one of the independent variables is completely randomized (different participants in each level) and the other is a repeated-measure (the same participant serves under each level of the other independent variable).

Repeated Measures ANOVA – can refer to either a one or more factor ANOVA where the same participant serves under every level of every factor.

Between-subjects – generally refers to an independent variable in the completely randomized ANOVA where the participants are randomly assigned to each level of a factor resulting in different participants in each level. It also refers to an *in situ* independent variable where different participants serve in each level of the variable.

Within-subjects – refers to an independent variable where the same group of participants serves under every level of the variable.

1. A researcher wishes to determine whether the controversial Healing Touch Therapy is successful when the patients are unaware that the therapist is using 'healing touch.' Twenty acute back pain patients are randomly assigned to either an Unaware condition where the patients sleep through the sessions or an Aware condition where they are awake. The patients then receive three sessions each of Healing Touch Therapy. After each session, the patients are asked to rate their pain on a 1 to 10 scale where 1 = no pain and 10 = maximum pain. Perform the appropriate statistical analysis. Does the design allow for the implication of causation?

	Unaware sessions				Aware sessions		
	1	2	3		1	2	3
Patient $_1$	8	8	9	Patient 11	5	3	2
Patient $_2$	7	7	8	Patient 12	6	7	4
Patient $_3$	5	4	3	Patient 13	5	3	2
Patient $_4$	4	6	4	Patient 14	4	3	1
Patient $_5$	7	9	6	Patient 15	4	2	1
Patient $_6$	5	5	4	Patient 16	3	1	1
Patient $_7$	8	9	0	Patient 17	3	3	3
Patient $_8$	10	9	9	Patient 18	4	3	2
Patient $_9$	9	8	9	Patient 19	2	4	2
Patient $_{10}$	8	9	9	Patient 20	2	2	1

REPEATED MEASURES ON BOTH FACTORS

2. An experimental neuropsychologist wishes to determine whether a new drug (X-9) enhances the growth of new, functional brain tissue after maximal brain growth has already been attained after a period of 1 month and 6 months. Five rats were run in a series of 3 different mazes. All of the mazes were judged to be of equal difficulty. The time in seconds to solve each maze is recorded for each rat across their 5 mazes. X-9 was given 2 weeks before testing. A prior study statistically demonstrated that rats not given X-9 did not improve their times. Is there any evidence X-9 works?

	1 month				6 months		
Trial	1	2	3	Trial	1	2	3
Rat 1	10	12	5		4	5	4
Rat 2	15	14	8		6	5	4
Rat 3	14	14	7		6	7	7
Rat 4	9	10	5		7	8	8
Rat 5	12	11	4		7	6	7

14 Nonparametric Statistics: Chi Square

ANOVA, *t* tests, and correlation are all examples of parametric statistics because the underlying distributions are assumed to be normal, and the distributions may be consequently described by common statistical parameters like the arithmetic mean, standard deviation, and variance. There are, however, many common situations where the underlying distribution is non-normal or unknown, or where the nature of the variable itself is neither ordinal, interval, or ratio. For these cases, statisticians developed a class of statistics known as **nonparametric** statistics. Nonparametric statistics are used when the data are measured by nominal or categorical scales or when the underlying distributions violate the assumptions of the parametric statistics. The most common use of nonparametric statistics is probably not the latter case but the former: when the data are in name only (like gender) and cannot be given any numerical value except arbitrarily, *e.g.*, 1 = male, 2 = female.

Nonparametric statistics are not the first choice of most statisticians. In nearly every experimental situation, a parametric statistic has more power (the ability to detect genuine relationships, genuine change or differences) than a nonparametric statistic. Yet some types of research, because of the nature of the data, must always be analyzed nonparametrically. The most popular of the nonparametric statistics is Chi Square or χ^2 (from the Greek letter χ, pronounced k-eye).

The Chi Square tests are used most often to analyze data that consists of counts or frequencies. For example, frequency data establishes how many people or events qualify for a particular category, like how many people choose Coke or Pepsi in a taste test. Notice that the data in a taste test ends up to be simple counts, like 110 people chose Coke and 123 chose Pepsi. Frequency data simply answer the question of 'how many?'

OVERVIEW OF THE PURPOSE OF CHI SQUARE

Chi Square statistics are designed to determine whether an observed number differs either from chance or from what was expected. For example, the simplest Chi Square design uses a single independent variable on a nominal or categorical scale with two levels, like gender. If we theorized that nothing other than pure chance should affect the gender of babies, then 50% of all babies should be male and 50% should be female. The experiment or study would consist of coding the gender of all births at a hospital or hospitals in a city, and a Chi Square test would determine whether the observed or actual birth rates differ from the theoretical expectation. Interestingly, in this example, they do. In the USA, male babies account for about 49% and female babies account for 51% of all births. When Karl Pearson created the Chi Square test in the 1890s, he termed it a **goodness-of-fit test** to describe how well theoretically the observed frequencies fit the expected frequencies.

OVERVIEW OF CHI SQUARE DESIGNS

The previous example is considered a two cell Chi Square design because there is only one independent variable with two levels. The levels are called **cells** in Chi Square analyses. Thus, as ANOVA was an extension of *t* tests (from a two group analysis to more than two groups to be analyzed), a two cell Chi Square can consist of a single independent variable with two cells or it can consist of more than two cells. Chi Square can also be similar to a factorial ANOVA, in that Chi Square can analyze two or more independent variables at once. The most commonly used Chi Square design is probably the 2×2 (pronounced two by two) design where there are two independent variables and each has two levels. To build on our previous example, we might test to see whether the birth ratio of males to females differs in America from India. One independent variable is gender (male or female) and the other is a geographical country (America or India). The null hypothesis states that these two independent variables are independent of each other (the ratio of births is the same in both countries even if the numbers are higher in India). The alternative hypothesis states that the two independent variables are dependent (which means that the ratios are different in the two countries, *i.e.*, the ratio of male babies to female babies is higher in India).

CHI SQUARE DESIGNS

Chi Square Test: Two Cell Design (Equal Probabilities Type)
The simplest form of a Chi Square test is a two cell experiment where the outcome of the experiment is that each participant or event falls into only one of the two cells. Let us continue with our gender examples. An experimenter has obtained 10 male and 15 female participants. The experimenter wishes to know whether there are significantly more female than male participants. If the participants were chosen from a population where the number of males and females is equal, then the probability of being male is 1/2 or .5, and the probability of being female is 1/2 or .5, and the sum of the two probabilities is 1.0.

COMPUTATION OF THE TWO CELL DESIGN
In a Chi Square test of this experiment the null and alternative hypotheses would be:

$H_0: P_M = P_F$

$H_a: P_M \neq P_F$

where

P_M = the probability of being male

and

P_F = the probability of being female

Step 1. Construct a table of the observed and expected cell frequencies.

	Male	Female
Observed cell frequency	10	15
Expected cell frequency	12.5	12.5

Note that the expected cell frequency is obtained by multiplying the theoretical probability of being male (1/2 or .5) times the total number of participants in the study (10 + 15 = 25). The expected cell frequency for both cells is .5 times 25 = 12.5. In this case, the expected cell frequencies are equal to each other because it was stated that way in the null hypothesis, hence, the title, 'Equal Probabilities Type.'

Step 2. The Chi Square value is obtained by the following formula:

$$\chi^2 = \Sigma \frac{(\text{Observed Cell Frequency}_i - \text{Expected Cell Frequency}_i)^2}{\text{Expected Cell Frequency}_i}$$

where

i = the first cell, then the second cell, etc. up to k, where k is the last cell

So,

$$\chi^2 = \Sigma \left(\frac{(10 - 12.5)^2}{12.5} - \frac{(15 - 12.5)^2}{12.5} \right)$$

$$\chi^2 = \Sigma \left(\frac{(-2.5)^2}{12.5} + \frac{(2.5)^2}{12.5} \right)$$

$$\chi^2 = \Sigma \left(\frac{6.25}{12.5} + \frac{6.25}{12.5} \right)$$

$$\chi^2 = \Sigma. (5 + .5)$$

$$\chi^2 = 1.00$$

Step 3. Next, the degrees of freedom (*df*) are obtained by the following formula:

df = Number of Cells – 1

$df = 2 - 1$

$df = 1$

Step 4. Compare the derived Chi Square value to the critical value in the Appendix containing the Chi Square distribution in the back of the book. The critical value of Chi Square at $p = .05$ with $df = 1$ is 3.84. The derived value of Chi Square = 1.00 does not exceed the tabled critical value of Chi Square = 3.84 at $p = .05$ with $df = 1$. Therefore, H_o is retained, and it is concluded that the observed cell frequencies do not significantly vary from the theoretical cell frequency beyond what might be expected by chance, $\chi^2(1) = 1.00$, $p > .05$. Thus, although it was observed that there were more females than males in the study, the differences were not statistically different beyond what might be expected by chance.

THE CHI SQUARE DISTRIBUTION

Like the *t* distribution or *F* distribution, the Chi Square distribution is actually a family of theoretical distributions. Like the *F* value in ANOVA, the Chi Square value is a one-tailed test but unlike the *F* distribution, Chi Square values have *df* based on the number of cells or categories and not on the number of observations or participants.

POWER IN CHI SQUARE ANALYSES

The Chi Square statistic possesses the same issue surrounding power that the parametric statistics do: It is that increasing the number of observations in a Chi Square design increases the likelihood of rejecting the null hypothesis and detecting real differences between the two cell frequencies. It is an issue because the same ratio of females to males (in the present example it is 1.5 or there are 1.5 times as many females as there are males) in this smaller sample does not reach statistical significance. However, in a larger sample, with the exact same ratio, the differences will be significant. In the present example, a total sample of 125 (50 males and 75 females) would have reached statistical significance. Chi Square designs can also suffer from an abuse of power by using far too many participants in order to make small differences between observed frequencies look like real ones instead of just chance differences.

ASSUMPTIONS OF THE CHI SQUARE TEST

Although it was stated that the Chi Square test makes no assumption about the shape of the underlying population, there are a few important assumptions when using the Chi Square test.

1. The scores in each cell should be independent of one another. This means that a score in one cell should have no effect on a score in another cell. A violation of this assumption could occur if participants became aware that Cell 1 was being chosen more than Cell 2, and as a result Cell 1 was chosen at even greater frequencies later in the experiment because the participants thought that the choice of Cell 1 was somehow better than Cell 2. Independence also means that each score (count or frequency) represents only one unique participant or event. The same participant may not be represented by more than one score.

2. There should be a minimum of 5 participants or events in any one cell. Violations of this assumption do not automatically disqualify the use of the Chi Square statistic, but we have seen that power is already reduced by the choice of a nonparametric statistic, and the Chi Square test requires substantial numbers of scores in order to be sensitive to real or genuine differences in the data.

3. The dependent variable in the Chi Square test is assumed to be a frequency or count, like numbers of participants. It is not appropriate to analyze continuous variables unless they have been dichotomized. One method of dichotomizing a continuous variable is to split the data into two halves by the median. Those above the median become Group 1 and those below the median Group 2. However, it would be unwise to use a nonparametric statistic on a dichotomized continuous variable if the distribution was normal or mound-shaped, since the equivalent parametric test would be much more powerful and sensitive to real differences in the data.

Chi Square Test: Two Cell Design (Different Probabilities Type)

There is a less common variation of the simple two cell Chi Square test where the theoretical probabilities are expected to be different for the two cells instead of being equal. Let us use the example of handedness. In the general population, 95% are right handed people and 5% are non-right handed people. An experimenter wishes to determine whether there are significantly more non-right handed people in a sample of 100 people with epilepsy, and the experimenter found 85 right handed people and 15 non-right handed people. Do these observed differences vary significantly from the theoretical expectation in the population?

COMPUTATION OF THE TWO CELL DESIGN

The null and alternative hypotheses are as follows:

H_o: P_R = .95 P_{NR} = .05

H_a: H_o is not true.

where

P_R = the probability of being a right handed person

and

P_{NR} = the probability of being non-right handed person

Step 1. Construct a table of the observed and expected cell frequencies.

	Right handed	Non-right handed
Observed cell frequencies	85	15
Expected cell frequencies	95	5

Note that the expected cell frequency is obtained by multiplying the theoretical probability of being right handed (95/100 or .95) times the total number of participants in the study (85 + 15 = 100). Thus, the expected cell frequency for the right handers is .95 × 100 = 95, and for non-right handers it is .05 times 100 = 5.

Step 2. The Chi Square value is obtained by the following formula:

$$\chi^2 = \Sigma \frac{(\text{Observed Cell Frequency}_i - \text{Expected Cell Frequency}_i)^2}{\text{Expected Cell Frequency}_i}$$

where

i = the first cell, then the second cell, etc. up to k,

where

k = the last cell

So,

$$\chi^2 = \Sigma \left(\frac{(85 - 95)^2}{95} + \frac{(15 - 5)^2}{5} \right)$$

$$\chi^2 = \Sigma \left(\frac{(-10)^2}{95} + \frac{(10)^2}{5} \right)$$

$$\chi^2 = \Sigma \left(\frac{100}{95} + \frac{100}{5} \right)$$

$$\chi^2 = \Sigma \ (1.05 + 20.0)$$

$$\chi^2 = 21.05$$

Step 3. Next, the degrees of freedom *(df)* are obtained by the following formula:

df = Number of Cells − 1

$df = 2 − 1$

$df = 1$

Step 4. The critical value of Chi Square is obtained from the Appendix containing the Chi Square distribution in the back of the book. The critical value of Chi Square at $p = .05$ with $df = 1$ is 3.84. The derived value of Chi Square = 21.05 does exceed this tabled critical value of Chi Square. Therefore, H_o is rejected, and it is concluded that the observed cell frequencies do significantly vary from the theoretical cell frequency beyond what might be expected by chance, $\chi^2(1) = 21.05, p < .01$. In terms of the research question, it was observed that there were significantly more non-right handed people in a sample of people with epilepsy than would be expected in the population of people without epilepsy.

INTERPRETING A SIGNIFICANT CHI SQUARE TEST FOR A NEWSPAPER

It is important throughout all of statistics to remind ourselves of the grand meaning of what we are doing. It is fine to report the study as significant, $\chi^2(1) = 21.05$, $p < .01$, to a sophisticated group of statisticians, but what would we tell our mothers or fathers we were doing (assuming they were not statisticians)? The contemporary psychologist-statistician Abelson asks his students, 'If your study were reported in the newspaper, what would the headline be?' (Abelson, 1995, pg. xiii).

I have long told my students to write a final paragraph of their study that summarizes the statistical findings in a way that the curious-but-unsophisticated reader could comprehend. In the previous example, it was concluded that there were significantly more non-right handed people in a sample of people with epilepsy than would be expected in people without epilepsy. Although the Chi Square test is not appropriate to analyze percentages, we could convert the frequencies in the cells into percentages after the Chi Square test is over in order to make grander sense of the data. For example, it was noted that 5% of the normal population is non-right handed. In this study, 15 people with epilepsy were non-right handed out of a total of 100. Thus, the percentage of non-right handed people with epilepsy was 15% (15/100). We could report to the newspapers or our parents that, while 5% of the normal population is non-right handed, it is significantly higher in people with epilepsy at 15%. Moreover, statisticians often divide the greater percentage by the smaller percentage and then report the rate of non-right handedness in people with epilepsy is three times (3×) higher than in people without epilepsy, and it is perfectly legal (after a significant Chi Square test) to state that there is 'a significantly higher rate of non-right handedness in people with epilepsy than in people without epilepsy.' While we did not directly test the significance of the percentages or the rates, it is completely valid to report on the statistical significance of these converted statistical parameters.

Chi Square Test: Three Cell Experiment (Equal Probabilities Type)

Another common form of a Chi Square test is a three cell design (and all of the following discussion and formulae can be applied to more than three cells) where the outcome of the experiment is that each observation (people or event) can fall into one of three cells. In this design, it is assumed at the outset that the probability of ending up in any cell is equal to the probability of ending up in any other cell. For example, an experimenter wishes to determine whether rats have a preference in a maze for a door on the right (D_1), a middle door (D_2), or a door on the left (D_3). Let us suppose that 30 rats ran the maze, and the observed frequencies for the three doors were: Door 1 = 13, Door 2 = 10, Door 3 = 7.

COMPUTATION OF THE THREE CELL DESIGN

In a Chi Square test of this experiment the null and alternative hypotheses would be:

$$H_o: P_{D1} = P_{D2} = P_{D3}$$

$$H_a: H_o \text{ is not true}$$

where

P_{D1} = the probability of choosing Door 1

and

P_{D2} = the probability of choosing Door 2

and

P_{D3} = the probability of choosing Door 3

Step 1. Construct a table of the observed and expected cell frequencies:

	Door 1	Door 2	Door 3
Observed cell frequencies	13	10	7
Expected cell frequencies	10	10	10

Note that the expected cell frequency is obtained by multiplying the theoretical probability of choosing Door 1 (1/3 or .33) times the total number of participants (rats) in the study (30). The expected cell frequency for all three cells is obtained the same way. The expected cell frequency is $1/3 \times 30 = 10$.

The Chi Square value is obtained by the following formula:

$$\chi^2 = \Sigma \frac{(\text{Observed Cell Frequency}_i - \text{Expected Cell Frequency}_i)^2}{\text{Expected Cell Frequency}_i}$$

where

$i =$ the first cell, then the second cell, etc. up to k,

where

$k =$ the last cell

So,

$$\chi^2 = \Sigma \left(\frac{(13-10)^2}{10} + \frac{(10-10)^2}{10} + \frac{(7-10)^2}{10} \right)$$

$$\chi^2 = \Sigma \left(\frac{(3)^2}{10} + \frac{(0)^2}{10} + \frac{(-3)^2}{10} \right)$$

$$\chi^2 = \Sigma \left(\frac{9}{10} + \frac{0}{10} + \frac{9}{10} \right)$$

$$\chi^2 = \Sigma (.90 + 0.0 + .90)$$

$$\chi^2 = \quad 1.80$$

Step 3. Next, calculate the degrees of freedom (df) by the following formula:

$df = $ Number of Cells $- 1$

$df = 3 - 1$

$df = 2$

Step 4. The critical value of Chi Square is obtained from the Chi Square distribution in the Appendix of this book. The critical value of Chi Square at $p = .05$ with $df = 2$ is 5.99. The derived value of Chi Square $= 1.80$ does not exceed the tabled critical value of Chi Square at $p = .05$ with $df = 2$. Therefore, H_o is retained, and it is concluded that the observed cell frequencies do not significantly vary from the theoretical cell frequencies beyond what might be expected by chance, $\chi^2 (2) = 1.80$, $p > .05$.

In terms of the research question, it appears that the rats have no greater preference for any door over any other door. Although it was observed that the door on the right was chosen with the greatest frequency, and the door on the left was chosen with the least frequency, the Chi Square test revealed that none of the doors was chosen with significantly greater or lesser frequency than what would be expected by chance.

CHI SQUARE TEST: TWO BY TWO DESIGN

A more sophisticated Chi Square design is the two by two design. It is actually used frequently in statistical analyses and is useful in many experimental situations. It is analogous to the two factor analysis of variance designs in some respects. The experimenter is interested in knowing whether two factors are independent of each other. If they are independent, then knowing a participant's score on one factor tells the experimenter nothing about that participant's score on the second factor. If the factors are dependent on one another, then knowledge about one factor can help predict the second factor. If the two independent variables are dependent and the null hypothesis has been rejected, then there is a significant and predictive relationship between the two variables.

For example, let us return to the factors of gender and handedness. Are these two factors independent of each other? Or would knowing a person's gender help predict their handedness? Let us suppose that a professor notes the frequencies of the gender and handedness of a 107 student psychology class. There are 22 right handed males, 5 non-right handed males, 76 right handed females, and 4 non-right handed females. Although it may not be obvious by observing the frequencies, the experimental question is whether the rate of non-right handedness in males is greater than the rate of non-right handedness in females. Thus, if the experimenter's hunch is true, and the two factors are dependent, then knowing a person's gender helps predict the likelihood of being right or non-right handed.

Computation of the Chi Square Test: Two by Two Design

The null and alternative hypotheses are as follows:

H_o: Gender and Handedness are Independent

H_a: Gender and Handedness are Dependent

Step 1. Construct a table of the observed cell frequencies.

	Male	Female		Total
Right handed	22	76	=	98
Non-right handed	5	4	=	9
Total	27	80	=	107

Step 2. Obtain the expected cell frequencies. For ease of computation, let us label the cells as follows:

	Male	Female		Total
Right handed	22 (a)	76 (b)	=	98
Non-right handed	5 (c)	4 (d)	=	9
Total	27	80	=	107

The expected frequency of Cell A is obtained by multiplying the column total for Cell A (27) times the row total for Cell A (98) and dividing by the overall total (107). Thus,

$$\text{Expected Cell A Frequency} = \frac{(27 \times 98)}{107}$$

Expected Cell A Frequency = 24.73

The expected frequency of Cell B is obtained by multiplying the column total for Cell B (80) times the row total for Cell B (98) and dividing by the overall total (107). Thus,

$$\text{Expected Cell B Frequency} = \frac{(80 \times 98)}{107}$$

Expected Cell B Frequency = 73.27

The expected frequency for Cell C is obtained by multiplying the column total for Cell C (27) times the row total for Cell C (9) and dividing by the overall total (107). Thus,

$$\text{Expected Cell C Frequency} = \frac{(27 \times 9)}{107}$$

Expected Cell C Frequency = 2.27

The expected frequency of Cell D is obtained by multiplying the column total for Cell D (80) times the row total for Cell D (9) and dividing by the overall total (107). Thus,

$$\text{Expected Cell D Frequency} = \frac{(80 \times 9)}{107}$$

Expected Cell D Frequency = 6.73

Now let us table the observed and expected cell frequencies:

	Male		Female	
	Observed	Expected	Observed	Expected
Right handed	22	24.73	76	73.27
Non-right handed	5	2.27	4	6.73

Step 3. The Chi Square test formula remains the same:

$$\chi^2 = \Sigma \frac{(\text{Observed Cell Frequency}_i - \text{Expected Cell Frequency}_i)^2}{\text{Expected Cell Frequency}_i}$$

where

i =the first cell, then the second cell, etc. up to k, where k is the last cell

So,

$$\chi^2 = \Sigma \left(\frac{(22 - 24.73)^2}{24.73} + \frac{(76 - 73.27)^2}{73.27} + \frac{(5 - 2.27)^2}{2.27} + \frac{(4 - 6.73)^2}{6.73} \right)$$

$$\chi^2 = \Sigma \left(\frac{(-2.73)^2}{24.73} + \frac{(2.73)^2}{73.27} + \frac{(2.73)^2}{2.27} + \frac{(-2.73)^2}{6.73} \right)$$

$$\chi^2 = \Sigma (.30 + .10 + 3.28 + 1.11)$$

$$\chi^2 = 4.79$$

Step 4. Next, the degrees of freedom (*df*) are obtained by the following formula:

df = (Number of Levels of Factor One – 1) × (Number of Levels of Factor Two – 1)

$df = (2 - 1) \times (2 - 1)$

$df = 1 \times 1$

$df = 1$

Step 5. The critical value of Chi Square is obtained from the Chi Square distribution in the Appendix. The critical value of Chi Square at $p = .05$ with $df = 1$ is 3.84. The derived value of Chi Square = 4.79 exceeds the tabled critical value of Chi Square = 3.84 at $p = .05$ with $df = 1$. Therefore, H_o is rejected, and it is concluded that the two factors, gender and handedness, are dependent, $\chi^2(1) = 4.79$, $p < .05$. In terms of the research question, it appears that in this sample, males have over a four times greater rate of non-right handedness (23% of the males were non-right handed) than do females (5%).

WHAT TO DO AFTER A CHI SQUARE TEST IS SIGNIFICANT

There is an equivalent to the multiple comparison tests of analysis of variance in Chi Square, and it is referred to as the **test of the standardized residuals**. This test examines each cell and determines whether the cell makes substantial contribution to the overall significance of the test. The formula, which is applied to each cell, is as follows:

$$R = \frac{\text{Observed Cell Frequency} - \text{Expected Cell Frequency}}{\text{Expected Cell Frequency}}$$

where

R = the absolute value of the standardized residual

Note that the absolute value of the standardized residual must exceed a value of 2.00 in order for that cell to be considered a major contributor to the overall significance of the Chi Square value. Interestingly, in the previous 2 × 2 example, none of the cells exceeds the critical value of 2.00. In this case, it appears that the test of standardized residuals is too conservative. Since post hoc analyses in Chi Square are relatively rare and in a theoretical sense all of the cells must contribute to the overall significance of the Chi Square statistic, one *post hoc* strategy (besides the test of standardized residuals) is to look simply at the individual Chi Square values of each cell in the original Chi Square formula or in the test of standardized residuals, and note their relative rankings. In the previous example's Step 3, it was observed that Cell C and Cell D contributed the most to the overall Chi Square value. Thus, it can be concluded that the relative *high* rate of non-right handed males and the relative low rate of non-right handed females had the most substantial contributions to the overall Chi Square's significance. It should also be noted that there will be a perfect correlation between the rankings of the cells in terms of their contributions to the original Chi Square formula and the rankings of the cells in the test of standardized residuals.

WHEN CELL FREQUENCIES ARE LESS THAN 5 REVISITED

Chi Square designs can become more complicated than a simple two by two design. There can be two or more levels of any number of factors such as 2 × 3, 2 × 10, or 2 × 2 × 3 designs. Many times in these more complicated designs, more than one cell might fall below the minimum observed cell frequency of 5. One solution in these designs is to collapse one of the factors into only two cells or three cells in order to increase the cell frequencies. Some information is invariably lost in this procedure, but the Chi Square test will be more sensitive to real differences among the cells with the increase in cell size.

For example, examine the following frequency table:

	\multicolumn{6}{c}{Handedness by education (in years)}
	9	10	11	12	Some college	BA or BS
Right	2	2	3	10	5	3
Non-right	1	1	1	3	3	2

In this 2 × 6 Chi Square design, only two of the 12 cells meet the minimum recommended observed cell frequency. However, if the Education Factor is collapsed into a dichotomous variable, such as 12th grade and below versus some college work or greater, then the following table results:

Handedness by education (in years)		
	12th grade or less	Some college or greater
Right	17	8
Non-right	6	5

Now, all of the cells meet the minimum criterion.

HISTORY TRIVIA

Karl Pearson (1857–1936) developed the Chi Square test in the 1890s. He termed it a test of goodness-of-fit by measuring the discrepancy between an actual distribution of numbers and a mathematical model based on probability theory. His son, Egon Pearson, later wrote that Chi Square was a 'new powerful weapon in the hands of one who sought to battle with the myths of a dogmatic world.' He also wrote that Chi Square was 'one of Pearson's greatest single contributions to statistical theory.' Helen Walker, a statistician and historian, wrote of Karl Pearson that few people '. . . in all of the history of science have stimulated so many other people to cultivate and enlarge the fields they have planted.'

In 1900, Karl Pearson wrote a paper on the similarities and differences in plants and submitted it to the Royal Society in England for publication. Pearson felt that the paper's referees were overly critical of his paper, and he thought that the mainstream biologists did not fully appreciate statistical methods. Although the paper did finally get published, an eminent Zoologist W.F.R. Weldon, who was sympathetic to Pearson's ideas, suggested that Pearson establish a journal devoted to the mathematical and statistical contributions to the biological sciences. This new journal was called *Biometrika*, and Pearson was its editor for almost 40 years.

KEY SYMBOLS AND TERMS

Nonparametric statistics – a variety of statistical tests that make no assumptions about the shape of the underlying population distribution. The word *nonparametric* implies that standard statistical parameters, like mean and standard deviation, do not apply or are not appropriate to these types of data.

Chi Square Test – one of the most popular nonparametric tests that involves the assessment of one or more independent variables, each with two or more levels of nominal or categorical data.

Goodness-of-Fit – Karl Pearson's original name for the Chi Square statistic. A Goodness-of-Fit test measures how well observed frequencies match theoretically expected frequencies.

Test of the Standardized Residuals – a post hoc analysis in Chi Square, analogous to multiple comparison tests in ANOVA, which determines which cells make substantial contributions to the overall significance of the Chi Square test.

1. The Humane Society has to set up a budget for the coming year. The concern is whether they need to buy more dog food or more cat food. Over the past year, they have housed 637 cats and 725 dogs. If the population of dogs and cats is theoretically equal, was the sample they cared for last year different from the norm?

2. A scientist decided to investigate whether tossing a coin really is a fair way to settle a bet. She went to the bank and got an uncirculated quarter. She tossed the quarter 100 times. The results were 56 heads and 44 tails. Is tossing this coin a fair way to settle a bet?

3. A study was done to determine if persons treated for an accident involving a joint injury were more likely to go to an orthopedist or a chiropractor. The local population has 26 chiropractors with approximately 65 patients each, while there are 18 orthopedists with approximately 70 patients each. Examining insurance company claim records, it was learned that 70% of people prefer a chiropractor to an orthopedist. Does the local population support the average?

4. A consumer psychologist was testing a new coffee blend. He randomly chose 230 people to try a taste of three samples (Sample A, Sample B, and Sample C). The results were that 98 people chose sample A, 42 people chose Sample B, and 90 chose Sample C. Were the three samples chosen equally?

5. A researcher wanted to know if babies prefer certain colors to others. Preference was measured by physical activity and pupil response. Seventy-two babies were placed in a baby seat and at different intervals were flashed dots of red, yellow, or blue. The preferences were 26 red, 22 yellow, and 24 blue. Was there a significant difference in preference?

6. A psychologist wishes to determine whether the diagnosis of Conduct Disorder is more prevalent in boys or girls. She analyzed consecutive admissions at a state mental hospital, and notes the following:

   500 Boys without Conduct Disorder

   265 Boys with Conduct Disorder

   387 Girls without Conduct Disorder

   56 Girls with Conduct Disorder

Develop a 2 x 2 Chi Square table and determine whether the diagnosis of Conduct Disorder is independent of gender.

7. A psychologist wanted to determine what was more important to the choice of a spouse: love or money. The psychologist surveyed 290 unmarried psychology majors and asked them to choose a prospective mate for either love or money. The psychologist was also interested to see if the choice would vary by gender. The psychologist found: 68 men preferred a spouse for love, 33 men preferred a spouse for money, 79 women preferred a spouse for love, and 110 women preferred a spouse for money. Are spouse choice and gender independent?

8. A psychologist wishes to determine whether Tourette's Syndrome occurs more in males than in females. She noted the gender and the absence and presence of Tourette's Syndrome in a large hospital's consecutive admissions for the past five years. Are gender and Tourette's Syndrome independent?

    1,024 Men without Tourette's

    27 Men with Tourette's

    986 Women without Tourette's

    9 Women with Tourette's

15 Other Statistical Parameters and Tests

There is a wealth of other statistical parameters, tests, and investigative procedures besides the ones presented in this book. Most scientific disciplines share a general body of statistical knowledge like basic statistical parameters such as the arithmetic mean, standard deviation, frequency distributions, hypothesis testing, and correlational procedures. Yet most disciplines also favor particular statistical tests and investigative procedures, and most have statistics that are unique to or favored by their discipline. The purpose of the present chapter is two-fold: first, to present an introduction to some of the statistical parameters that are frequently used in medicine, nursing, epidemiology, and health science disciplines and, second, to present an overview of some of the more popular yet more complicated statistical tests and procedures, known as **multivariate statistics**. Multivariate statistical procedures have formulas so complicated that their computation can only be performed by computers, and their interpretation is not without controversy.

HEALTH SCIENCE STATISTICS

Test Characteristics

While the health sciences share most of the statistical parameters, tests, and experimental investigative procedures presented in this book, these sciences also employ some discipline-specific parameters. Many of them come from the discipline of epidemiology, which is the study of disease. One of the most essential concepts in this discipline is also part of signal detection theory which was discussed earlier in this book.

Signal detection theory is reconfigured in epidemiology and can be graphically represented by a **receiver operating curve** (ROC). Before panic sets in, allow me to explain at the outset that these are relatively simple ideas, parts of which you have undoubtedly used or have understood intuitively already in your lives. At its heart, we are concerned with whether a disease is truly present or absent. For example, the early detection of cancer is extremely important to us all, since cancer treatment remains as primitive as ever: burn it, poison it, or cut it out. And statistics reveal that the earliest treatment of cancer is associated with a better prognosis. Many types of cancers have tests which are designed to predict their earliest presence or precancerous conditions. However, the tests are not perfect. Ideally, we would prefer that the test works perfectly and is never wrong.

There are two ways a test can be right and two ways a test can be wrong. One way for the test to be right is for the test to indicate the presence of cancer only when the patient really does have cancer. This situation is said to be a **true positive**. The first word in this tandem refers to the genuine truth or falsity of the test's decision. The second word refers to whether the test indicated the presence of the disease (positive) or absence (negative) of the disease. A true positive indicates that the test determined that cancer was present (the test results were positive) and, indeed, the patient really did have cancer (true positive). The other way the test can be right is by a **true negative**. In this situation, the test has determined that the patient does not have cancer (the test results were negative), and it is subsequently found that the patient did not have cancer (true negative).

One way the test can be wrong is by a **false positive**. A false positive indicates that the test determined that cancer was present; that is, the test came out positive, but the test was wrong, and the patient did not really have cancer. The other way the test can be wrong is by a **false negative**. In this situation, the test indicates that the patient does not have cancer (the test results are negative) when the patient really does have cancer (false negative). While the latter mistake has mortal consequences, so may the former mistake. With a false negative, the cancer may spread until the cancer cannot be treated because the test failed to pick up the earliest signs of the disease. We would always hope that the test we use to detect the earliest signs of cancer would be **sensitive** to the earliest signs of cancer. Thus, a characteristic of a good test is the test's **sensitivity**.

A test possesses good sensitivity if it has a high true positive rate. In practice, this rate may come at some cost, and that is, it may also have a few false positives in order to achieve a high true positive rate. In other words, the test is a good one if it identifies nearly all of the people who really have the disease, plus a few people who do not. While a false positive indication does not have all of the dire consequences of a false negative, there are consequences: A patient with a false positive indication may unnecessarily be subjected to unneeded treatment or surgery. The patient may even be harmed or killed as a result of these treatments. The adverse effects of the treatment of a disease are referred to as **iatrogenic effects**. With the disease of cancer, either false positives or false negatives can have mortal consequences because some treatments of cancer have mortality rates associated with them. Refer to Figure 15.1 for a summary of these decisions.

This model of disease detection does assume that the patients have been followed-up long term and, in the final analysis, there is no question as to whether the patient did or did not have the disease. This assumption is obviously questionable, and we will discuss later in the chapter how test results and outcomes may be specious. Nevertheless, epidemiologists use the model to explain the characteristics of good and bad tests, and the model is certainly useful in disease prevention and treatment.

As discussed earlier, a test's sensitivity is defined as its ability to detect most

		Has the disease	Does not have the disease
This test indicates:	Positive	True Positive a	False Positive b
	Negative	False Negative c	True Negative d

Figure 15.1

of the people with the disease and maybe a few patients who do not have the disease. Sensitivity may be calculated as a proportion by taking the number of true positives and dividing by the total number of people with the disease. Thus,

$$\textbf{Sensitivity} = \frac{a}{a + c}$$

If a test is reported to have a sensitivity of 87%, it means that of 100 patients who are really known to have the disease, 87 of them tested positively for the presence of the disease (true positives). The other 13 are considered to be false negatives because the test said they do not have the disease, but they do.

Another characteristic of a good test is its **specificity**, and this means that it should also correctly rule out people who do not have the disease, although it may include a few people who do not have the disease but the test said they do (false positives). A test's specificity may be calculated as follows:

$$\textbf{Specificity} = \frac{d}{b + d}$$

Another way of thinking about a test's specificity is to answer this question: If a person does not have the disease, what is the probability the test will be negative? If a test has a specificity of 79%, it means that out of 100 people who are really known not to have the disease, 79 of them did not have the disease according to the test (true negatives). The other 21 are considered to be false positives because the test said they have the disease, but they do not.

The overall **diagnostic accuracy** may be calculated by combining the total number correct (true positives + true negatives) and dividing by the total combination of all outcomes (true positives + false positives + true negatives + false negatives). The diagnostic accuracy of a test is sometimes called the **efficiency** of a diagnostic test. Hence,

$$\text{Diagnostic Accuracy} = \frac{a + d}{a + b + c + d}$$

Another increasingly popular measure of diagnostic accuracy is the likelihood ratio. The **likelihood ratio for a positive test** combines the sensitivity and specificity of a test into a single parameter. This ratio expresses the likelihood that the test will be positive in a person with the disease compared to a positive test in a person without the disease. The formula consists of dividing the test's sensitivity by 1 minus the test's specificity. Thus,

$$\text{Likelihood Ratio for a Positive Test} = \frac{a/(a + c)}{1 - (d/(b + d)}$$

The likelihood ratio of a positive test might be reported as 8.5, which would be interpreted that a positive result of the test is 8.5 times more likely to be seen in a person with the disease compared to a person without the disease.

The prevalence of a disease may also be derived by the previous characteristics. Prevalence of a disease is defined as the number of people in a population who have the disease at a particular point in time divided by the total number of people in the entire population (those with and without the disease). Thus,

$$\text{Prevalence} = \frac{a + c}{a + b + c + d}$$

The **incidence** of a disease is defined as the number of verified new cases of the disease in a population for a period of time (usually one year) divided by the total number of people in the population. The parameters of prevalence and incidence are useful in the prediction of diseases and helpful in decision-making regarding the control and treatment of diseases. For example, although autistic disorders in children are well known and stories about the disease are common in the media, the prevalence of the disease is actually very low and its incidence is extremely stable. In contrast, the prevalence of AIDS in some countries used to be the same as autistic disorders but the incidence has increased dramatically every year. In the same fashion, in some countries while the prevalence of AIDS reached higher and higher levels, the incidence began to drop. Some epidemiologists inferred that the drop in incidence indicated that prevention programs had begun to work.

Earlier it was noted that sometimes test results are not so clear, and sometimes the final determination may even be uncertain or equivocal. One common example of **equivocal results** comes from home pregnancy tests. In some of these kits, the test is supposed to turn red if the woman is pregnant or stay white if she is not pregnant. But what happens if the indicator turns pink? This is an example of an **intermediate result** where the result is neither positive nor negative but falls between these two poles.

A second type of equivocal result is the **indeterminate result**. This means that the test result is neither positive nor negative and does not fall between the two poles. In this case, there may be a variety of reasons related or unrelated to the test that precludes a valid reading. An example of an indeterminate result is when an x-ray cannot determine whether a person has a bone fracture or not, or an MRI cannot tell whether a person has sustained a closed head injury.

A third type of equivocal result is the **uninterpretable result** where the test was not performed according to the standardized procedures. Uninterpretable results can occur because the person using the test is not completely familiar with the testing procedures. Equivocal results may be annoying but they are common throughout the research realm. It may or may not be the fault of the experimenter. In cases where it may not be the experimenter's fault, the test may be state-of-the-art, but it may also be that the 'art' is simply not that good at this point in time. In the latter case, it would behoove anyone involved with a scientific experiment to be fully informed of the experimental and standardized procedures. In dealing with equivocal results, it is the responsibility of the researcher to report the frequency or proportion of equivocal results, since its publication may motivate others to improve upon inadequate tests, improve upon testing procedures, etc.

RISK ASSESSMENT

A common method of determining which procedure should be chosen in health related professions is the statistical assessment of risk. As noted earlier, diagnostic procedures and treatments may themselves have harmful or even deadly (iatrogenic) effects, so their choice becomes a very important decision. **Absolute risk reduction** refers to the reduction in risk between two groups treated differently when their outcomes are compared: For example: Imagine if, after a first heart attack, 100 women took vitamin E supplements daily for three years while 100 similar women did not. The incidence of a second heart attack is determined for both groups over the three-year period. If the incidence of a second heart attack in the vitamin E group is 11% and the incidence for the unsupplemented group is 14%, then the absolute risk reduction is 3% (or 14%–11%). The absolute risk reduction is formulaically obtained by subtracting the proportion of people with the outcome of interest who were untreated from the proportion of those who were treated. Thus,

Absolute Risk Reduction = Proportion of Treated Group's Outcome – Proportion of Untreated Group's Outcome

An absolute risk reduction of 3% indicates that vitamin E supplements may reduce the risk of heart attack by 3%. It is a separate statistical procedure, significance testing, which determines whether these results could have occurred by chance and, even under those circumstances, replication of the findings by other experimenters would be necessary for the findings to be trustworthy. Absolute risk reduction is often coupled with significance testing, making it a highly valuable and easily interpreted parameter of general statistical interest.

The relative risk reduction is a similar parameter except it is expressed as a percentage. **Relative risk reduction** is a measure of how much reduction in risk (as a percent) has occurred between two groups treated differently (most commonly, the experimental and control groups) on an outcome of interest when compared to the untreated group's percentage of the outcome of interest. The formula is:

Relative Risk Reduction $= [(P_C - P_E)/P_C] \times 100\%$

where

$P_C =$ the proportion of the control group with the outcome of interest

$P_E =$ the proportion of the experimental group with the outcome of interest

In a recent study of 1,100 potato chip eaters, experimenters were interested in the incidence of gastric troubles and flatulence in a diet potato chip compared to regular potato chips. It was found that the incidence (of these outcomes for 48 hours) for the diet chip was 16% while the incidence was 18% for the regular chip. Thus, the relative risk reduction (RRR) would be calculated as RRR = $[(18\% - 16\%)/18\%] \times 100\%$ or 11%. The experimenters observed an 11% reduction in gastric problems and flatulence with the new diet potato chips compared to regular potato chips. Once again, however, only significance testing would reveal whether the observed relative risk reduction was attributable to chance.

PARAMETERS OF MORTALITY AND MORBIDITY

Mortality refers to the proportion of deaths relative to the population while morbidity refers to the incidence of disease or sickness in a sample or population. Since the specific causes of death (morbidity) are more difficult to measure than death itself (mortality), there are more indicators of mortality than morbidity. Nevertheless, there are a wide array of rates, proportions, and percentages based on mortality and morbidity throughout the health sciences. One of the simplest and most common measures of the general health of a population is the annual crude death rate. The **annual crude death rate** is defined as the number of deaths in a year divided by the population (as of mid-year or July 1 of that same year). Frequently, the quotient in this determination is multiplied by a factor of 100 or a higher multiple (1,000, 1,000,000, etc) in order to make more interpretable sense of the statistic. The multiple is chosen based on the magnitude of the quotient. If the quotient is very small (*e.g.*, .000001) it would be senseless to use a base of 1,000 while with a larger quotient (.067) it might make perfect sense to use a base of 1,000. The formula is:

Annual Crude Death Rate

$$= \frac{\text{Number of Deaths in a Calendar Year}}{\text{The Population at Mid-Year}} \times \text{A Multiple of 100 or Higher}$$

For example, approximately 2,144,000 died in the U.S.A. last year. The population at mid-year was approximately 263,000,000 (these two numbers are not official estimates). So the calculation would be:

$$\textbf{Annual Crude Death Rate} = \frac{2,144,000}{263,000,000} \times 1,000$$

Annual Crude Death Rate = 8.0 deaths per 1,000 people

The annual crude death rate is used frequently to compare the general health of countries across the world. It is acknowledged that it is not an exact indicator of the health status of a country, since it sums across such factors as age, gender, ethnicity, and conditions such as war and natural disasters. In 1992, the annual crude death rate in Costa Rica was 4.0 deaths per 1,000, one of the lowest in the world, while in Afghanistan the annual crude death rate was 22.1 deaths per 1,000, one of the highest in the world.

Death rates may also be reported as age-specific, cause-specific, age-cause specific or other combinations. In **age-specific death rates**, an age range is specified (*e.g.*, 16–24 years) and the number of deaths, in that age range (for a given year), is divided by the total population of that age range at mid-year. Factor-specific death rates are useful in determining populations of people at risk as a function of their age, gender, ethnicity or other variables and are very frequently used as arguments before political groups for health intervention and disease prevention. In cause-specific death rates, a cause of death (*e.g.*, motor vehicle, accidents, suicide, etc) is specified, and the number of deaths due to that specific cause is divided by the population at mid-year. For example, motor vehicle accidents have the greatest cause-specific death rates of all accidental deaths in the U.S.A. at approximately 17 deaths per 100,000.

In the prevention and treatment of disease, the incidence of morbidity is a highly useful parameter even though morbidity can be much more difficult to determine than mortality. Morbidity is concerned with the measurement of disease, and many times the stigma and biases associated with diseases prevent their accurate report. For example, Freddy Mercury, the lead singer of the rock group Queen, publicly admitted he had AIDS the day before he died of the disease.

Probably the most common measure of morbidity is incidence rate. **Incidence rate** is defined as the number of new cases of a particular disease in a year divided by the population at mid-year. The quotient is multiplied by an interpretable factor like 1,000 or higher. For example, in 1992 there were 28,215 new cases of AIDS reported. The total population of the U.S.A. at that time was about 257,000,000. Thus, the incidence of AIDS in 1992 was 11.0 cases per 100,000.

The prevalence proportion is another useful statistical parameter associated with morbidity. The **prevalence proportion** is defined as the total number of cases of a disease known to exist at a given point in time divided by the total population at the same time. This quotient is also multiplied by an interpretable multiple like 1,000, 100,000, etc. The prevalence proportion can also be reported specific to an age, gender, or other demographic combination. For example, in 1992 there were 1,991 known cases of AIDS in Denver, CO. The population in Denver at that time was about 1,700,000. The prevalence proportion was 11.7 per 10,000. The prevalence proportion of AIDS in Miami, FL at this same time was 38.4 per 10,000.

The case-fatality proportion is used as an indicator of how serious a particular disease might be. The **case-fatality proportion** is defined as the number of deaths attributable to a particular disease during a period of time divided by the prevalence proportion of the disease at the same time. The quotient is reported as an interpretable multiple. For example, in 1992 there were 23,411 known deaths from AIDS in the U.S.A. The prevalence of AIDS patients at that time was approximately 82,000. Thus, the case-fatality proportion was 28.5 per 100 cases. Obviously, the latter figure would indicate that AIDS was a very serious and fatal disease in 1992. With the advent of new drug combinations for the treatment of AIDS, the case-fatality proportion for people with AIDS has fallen dramatically. Despite its usefulness and popularity, the case-fatality proportion is still fraught with interpretation and measurement problems. Because AIDS, in particular, is known to mutate, it is possible that the newer cases of the disease may not have the same mortality rates as earlier cases of the disease. Also, because of successful new treatments of AIDS, newer cases may have lower mortality rates than older or more advanced cases. At best, the case-fatality proportion is a useful but crude index of the seriousness of a disease.

MULTIVARIATE STATISTICS

With the advent of computers, the field of statistics was vastly advanced in an area known as multivariate statistics. Multivariate statistics may have similar experimental designs to those previously mentioned throughout this book but they are concerned with the analysis of more than one dependent variable at one time. The Pearson correlation coefficient could be considered the simplest example of a multivariate statistic, since the correlation coefficient analyzed two different continuous variables at once. However, it is traditionally considered a bivariate statistic rather than a multivariate statistic. Another statistic in this debatable category is an analysis of covariance.

ANALYSIS OF COVARIANCE

An **analysis of covariance** (ANCOVA) is a method of determining differences among means, on a single dependent variable, for the levels of an independent variable. ANCOVA can have the same designs as in ANOVA; however, an ANCOVA statistically controls for a confounding or nuisance variable. For example, imagine a researcher was interested in age differences in working memory (short-term memory). Fifteen younger people and 15 older people were tested for their working memory (the maximum number of digits that could be repeated in a listening paradigm). Consistent with the researcher's hypothesis, the younger people showed a small but statistically significant advantage. During the preparation of the manuscript for publication, the researcher reads that working memory appears to have a strong positive correlation with IQ ($r = .50$). Perhaps the younger people in the study simply have higher IQs than the older people and IQ not age, accounts for the difference in working memory. The researcher happens to have an estimate of the participants' IQs because of an elaborate pre-testing procedure. An ANCOVA will remove or covary the effect of the relationship of IQ upon working memory, and still allow for the assessment of the effect of age upon working memory.

ANCOVA has all of the assumptions of ANOVA and, in addition, ANCOVA assumes that the effect of the covariate (the nuisance variable) is the same on the dependent variable under all levels of the independent variable. It is also important to note that the covariate should have been measured prior to the measurement of the dependent variable. A violation of this latter assumption may mean that the ANCOVA will remove important and useful variance from the dependent variable if they are gathered at the same time.

MULTIVARIATE ANALYSIS OF VARIANCE

A **multivariate analysis of variance** (MANOVA) is similar to all design forms of ANOVA with the exception that two or more dependent variables are measured at once. A MANOVA can give a better picture of the complex effects of independent variables upon behavior, since it allows for the measurement of more than just one dependent variable. However, the cost of this picture is complexity both in computation (it can only be performed by computer) and in interpretation. When a simple one factor MANOVA is statistically significant, it indicates that the combination of dependent variables varies among the levels of the independent variable. The interpretation, however, is restricted to univariate post hoc interpretations of each dependent variable by itself. A significant MANOVA does not reveal the pattern of significant differences.

Furthermore, there is a controversy about whether a MANOVA provides protection against the Type I Error when using multiple dependent variables. It has been often argued that it does provide protection against the Type I Error. Whether it is true or not, the inclusion of a number of dependent variables in a MANOVA that are not thought to vary as a function of the independent variable may increase the probability of the Type II Error. In this situation, the combination of irrelevant dependent variables hinders the detection of a potentially significant dependent variable. One moral of this tale is that one should always be thoroughly familiar with a statistical procedure, able to defend its use, and know of its abuses. Otherwise, it is better to employ statistical procedures with which one is comfortable. The ultimate criterion on a statistical choice should be whether the statistic helps the experimenter to understand the phenomenon in question better.

MULTIVARIATE ANALYSIS OF COVARIANCE

A **multivariate analysis of covariance** (MANCOVA) consists of analyzing the effects of one or more independent variables upon more than one dependent variable while controlling for the effect of one or more covariates (nuisance or confounding variables) upon these dependent variables. Again, the interpretation of a MANCOVA can be difficult. One of my last classes in graduate school was a course in multivariate statistics, and I used MANCOVAs in my dissertation. During my final oral defense, one member of my dissertation examining committee sat unusually quiet. After my successful defense, I asked him why he asked me no questions. 'Because I didn't understand a darn thing you did,' was his reply.

FACTOR ANALYSIS

There are a number of purposes of the family of statistical techniques known as **factor analysis**. Factor analysis consists of determining what an underlying conceptual structure looks like in a set of dependent variables by examining the correlations between each variable in the set with every other variable in the set. Factor analysis helps to determine which variables in a set are highly correlated. Thus, factor analysis can be used to reduce a set of variables to a smaller set by removing redundant variables. Factor analysis can also be used to identify underlying factors in a large set of variables so that the entire set may be better understood conceptually. For example, factor analyses on a set of differing intellectual tasks yielded a single underlying factor: *i.e.*, most of the variables in the set of tasks were intercorrelated and all were defined by a single underlying factor. This factor was subsequently named the '*g*' factor for general intelligence.

The experimental design in a factor analysis does not consist of levels of an independent variable. There is only one large sample of participants measured on a number of dependent variables. It is recommended that the number of participants should be greater than 10 times the number of dependent variables. If there are 15 items, questions, or dependent variables then there should be at least 150 participants in the factor analysis in order to yield replicable results.

For example, one concern in contemporary personality research is the search for the core number of factors required to describe the normal and abnormal personality. Factor analysis is the primary statistical technique that has been used to answer this question. Early factor analytic attempts to solve this question employed hundreds of behavior measures in order to determine how these traits related to each other and to determine whether there might exist a basic number of factors to which all other human traits might be related. Factor analyses, of course, are dependent themselves upon which dependent variables are included in the original analysis. Current researchers argue that three, five, seven, or sixteen core factors underlie personality. One very interesting aspect of factor analysis is that there is an infinite number of correct solutions.

MULTIPLE REGRESSION

Multiple regression is an extension of bivariate correlation. The word 'regression' generally implies a concern with prediction while the word 'correlation' implies that the concern is with the relationship among variables. In **multiple regression**, a single dependent variable (or criterion variable) is predicted from several independent variables (or predictor variables). The research concern is often two-fold: to what extent can the independent variables predict the dependent variable, and what is the strength of each independent variable in the prediction of the dependent variable. There are many different types of multiple regression, like step-wise regression, forward regression, and backward regression. All of these techniques vary the way in which the independent variables are introduced into the regression equation. The different techniques may give different pictures of the interrelationships between the independent variables and the dependent variable.

For example, a large sample of first grade children could be tested for their IQ and a large number of demographic variables might be collected at the same time. The interest would be to what extent can IQ (the dependent variable) be predicted from independent variables such as the socioeconomic status of the child's family, the child's mother's IQ, the child's father's IQ, the number of years of education of the parents combined, etc. Multiple regression would yield parameters that would describe the strength of the relationship between the criterion variable and the predictor variables and how much of the total variance in the child's IQ score could be accounted for by the predictor variables. The multiple regression equation would also yield weights that would reveal which was the strongest and weakest of the predictor variables in terms of their relationship to the criterion variable. Again, it is recommended that the total number of participants be at least 10 times the number of independent variables, and the results will be limited to the independent variables used and the sample upon which the variables were tested. There could be variables that might be more predictive of IQ but if they were not included in the multiple regression equation, their value remains unknown.

CANONICAL CORRELATION

Canonical correlation is essentially a correlation-regression technique in combination with a kind of factor analysis that yields different estimates of how two different sets of variables may be related to each other. The two sets of variables are specified as the criterion set and the predictor set, and the experimental question is to what extent is a weighted combination of the criterion set correlated with a weighted combination of the predictor set. The resulting correlation is called the canonical correlation coefficient. A canonical analysis also derives an amount of variance that is accounted for in the criterion set by the predictor set. While canonical correlation is not exceptionally popular, it is useful in determining complex relationships (as in a test's construct validity) between two psychological tests.

LINEAR DISCRIMINANT FUNCTION ANALYSIS

Multiple regression was used to determine how a criterion variable could be predicted by a set of independent variables or predictor variables. **Linear discriminant function analysis** (LDFA) is similar in that independent variables are used to predict group membership. In this design, the criterion variable is dichotomous instead of continuous. There is also a family of LDFA techniques that differ on the entry of the predictor variables in the LDFA equation.

LDFA has been used extensively in neuropsychology to predict whether a patient's set of neuropsychological test scores are consistent with those of patients with known neurological damage or those of patients without neurological damage. Note that LDFA requires that the two criterion groups are specified before the LDFA. Their differences (as in brain-damaged or not brain-damaged) are an *a priori* assumption. The subsequent LDFA will decide the extent to which these groups may be successfully classified according to the predictor variables. One problem associated with LDFA is the problem of shrinkage. Shrinkage refers to the phenomenon where a LDFA equation, derived on a sample of data, will have substantially less predictive power when applied to a new sample. This is often seen where LDFA on a specific sample has a high classification rate (usually around 90% or better), yet falls near 50% when the same LDFA equation is used to classify another sample.

CLUSTER ANALYSIS

The object of **cluster analysis** is to classify a large set of objects (people, plants, etc.) into distinct subgroups based on predictor variables. These subgroups will be homogeneous with respect to the group's scores on the predictor variables if the cluster analysis is successful. Cluster analysis is also referred to as hierarchical cluster analysis and taxonomy analysis. The statistical object of cluster analysis is to minimize the within-group variation in each group while maximizing the between-group variation among the groups. The cluster analyses' success will be limited by the ability of the predictor variables to yield identifiable clusters. Most clustering procedures also yield coefficients of group similarity that give relative estimates of how alike the members of each group are. A cluster analysis might show that Group 1, consisting of Participants 1, 7, 23, 33, 39, 42, 67, and 88, has a group similarity of $r = .58$ while Group 2, consisting of Participants 2, 3, 5, 6, 19, and 55, had a similarity profile of $r = .89$. Examination of graphic plots of each of the groups of the participants' means on the predictor variables allows a labeling of the groups such as the low achievers, over achievers, medium achievers, etc., depending on the nature of the predictor variables. Hierarchical cluster analysis allows the researcher to determine what number of groups is optimal since the procedure continues to combine the objects until there are only two groups (but the profiles of group similarity will be low and sometimes senseless at that point).

A SUMMARY OF MULTIVARIATE STATISTICS

As Abelson has noted, 'Don't use Greek, if you don't speak the language.' While multivariate techniques are impressive, their interpretation can be complex and controversial. Anyone who chooses to perform a multivariate statistical analysis should understand its assumptions, understand the repercussions of the violations of the assumptions, understand the computer print-out, and be able to defend the choice of the statistic. As another contemporary statistician, Nunnally, wrote, 'Don't let the tail wag the dog.' If univariate statistics will answer your experimental question just as well or if you will be able to understand and explain a univariate analysis of your data better than a multivariate statistic, then your choice is clear. One should also not shy away from new techniques, as they may yield new and interesting relationships in your data heretofore unrecognized. Also, remember that few people in the whole world are fully and completely comfortable with their knowledge of statistics. Statisticians and researchers are continually contacting one another asking for advice and guidance. So do not hesitate, during the rest of your statistical career, to seek help, advice, and guidance on the choice of your experimental design and the choice of your statistical analyses. And still be guided by your ever-growing statistical intuition.

Multivariate Statistics – a family of different statistical designs and procedures involving the analysis of more than one dependent variable at the same time. Multivariate statistics are performed only by computer programs, and their interpretation can be complex and often controversial.

Receiver Operating Curve – a graphical representation of signal detection theory where the sensitivity and specificity of tests can be visually analyzed.

True Positive – the condition where a test indicates the presence of a disease, and the patient really does have the disease.

True Negative – the condition where a test does not indicate the presence of a disease, and the patient really does not have the disease.

False Positive – the condition where a test indicates the presence of a disease, and the patient really does not have the disease.

False Negative – the condition where a test does not indicate the presence of a disease, and the patient really does have the disease.

Sensitivity – a characteristic of a good test where it is very likely to indicate the presence of a disease when the patient really does have the disease.

Specificity – a characteristic of a good test where it is very likely to be correct when it does not indicate the presence of a disease.

Diagnostic Accuracy – is represented as a proportion of the number correct (true positive + true negatives) divided by the total combination of all outcomes (true positives + true negatives + false positives + false negatives).

Likelihood Ratio for a Positive Test – combines the sensitivity and specificity of a test into a single parameter. The ratio expresses the likelihood that the test will be positive in a person with the disease compared to a positive test in a person without the disease.

Prevalence – is the number of people who have a specified disease at a given point in time divided by the total number of people in the population (with and without the disease).

Incidence – is the number of verified new cases of a disease in a period of time (usually one year) divided by the total number of people in the population (with and without the disease).

Equivocal Results – a condition when test results are unclear, uncertain, or unreadable.

Intermediate Results – occurs when a test result is neither positive or negative but falls between these two conditions.

Indeterminate Results – occurs when a test result is neither positive nor negative and does not fall between these two conditions.

Uninterpretable Results – occurs when a test has not been given according to its correct instructions or directions.

Absolute Risk Reduction – is the reduction in risk between two groups treated differently when their outcomes are compared.

Relative Risk Reduction – is a measure of how much reduction in risk has occurred between two groups treated differently when compared to the untreated group's percentage on the outcome of interest.

Annual Crude Death Rate – is the number of deaths in a year divided by the population as of mid-year of that same year.

Age-Specific Death Rate – is the number of deaths in that given age range (for a given year) divided by the total population of that age range at mid-year.

Prevalence Proportion (of a disease) – is defined as the total number of cases of a disease known to exist at a given point in time divided by the total population at the same time. This quotient is also multiplied by an interpretable multiple like 1,000.

Case-Fatality Proportion – is the number of deaths attributable to a particular disease during a period of time divided by the prevalence proportion of the disease at the same time. The quotient is reported as an interpretable multiple like 1,000.

Analysis of Covariance (ANCOVA) – is a method of determining differences among means, on a single dependent variable, while statistically adjusting for the effects of a confounding or nuisance variable.

Multivariate Analysis of Variance (MANOVA) – is similar to all design forms of ANOVA with the exception that two or more dependent variables are measured at once. The interest is whether there are group differences in the means on the dependent variables.

Multivariate Analysis of Covariance (MANCOVA) – consists of analyzing the effects of one or more independent variables upon more than one dependent variable while controlling for the effect of one or more covariates (nuisance or confounding variables).

Multiple Regression – a single dependent variable (or criterion variable) is predicted from several independent variables (or predictor variables). The research concern is often two-fold: to what extent can the independent variables predict the dependent variable, and what is the strength of each independent variable in the prediction of the dependent variable?

Canonical Correlation – yields estimates of how two different sets of variables may be related to each other. The two sets of variables are specified as the criterion set and the predictor set, and the experimental question is to what extent is a weighted combination of the criterion set correlated with a weighted combination of the predictor set.

Linear Discriminant Function Analysis (LDFA) – uses a set of predictor or independent variables in order to predict group membership on a dichotomous criterion variable. The interest is in whether the predictor variables can accurately classify the *a priori* specified dichotomous criterion variable.

Cluster Analysis – classifies a large set of objects into distinct subgroups based on predictor variables. These subgroups will be homogeneous with respect to the group's scores on the predictor variables.

Factor Analysis – a family of statistical techniques that determines the underlying conceptual structure of a set of variables or it can be used to reduce a set of variables by eliminating overlapping items.

Appendix A
The Z Distribution

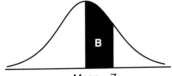

Mean Z

(A) z	(B) Area Between Mean and z	(C) Area Beyond z	(A) z	(B) Area Between Mean and z	(C) Area Beyond z	(A) z	(B) Area Between Mean and z	(C) Area Beyond z
.00	.0000	.5000	.45	.1736	.3264	.90	.3159	.1841
.01	.0040	.4960	.46	.1772	.3228	.91	.3186	.1814
.02	.0080	.4920	.47	.1808	.3192	.92	.3212	.1788
.03	.0120	.4880	.48	.1844	.3156	.93	.3238	.1762
.04	.0160	.4840	.49	.1879	.3121	.94	.3264	.1736
.05	.0199	.4801	.50	.1915	.3085	.95	.3289	.1711
.06	.0239	.4761	.51	.1950	.3050	.96	.3315	.1685
.07	.0279	.4721	.52	.1985	.3015	.97	.3340	.1660
.08	.0319	.4681	.53	.2019	.2981	.98	.3365	.1635
.09	.0359	.4641	.54	.2054	.2946	.99	.3389	.1611
.10	.0398	.4602	.55	.2088	.2912	1.00	.3413	.1587
.11	.0438	.4562	.56	.2123	.2877	1.01	.3438	.1562
.12	.0478	.4522	.57	.2157	.2843	1.02	.3461	.1539
.13	.0517	.4483	.58	.2190	.2810	1.03	.3485	.1515
.14	.0557	.4443	.59	.2224	.2776	1.04	.3508	.1492
.15	.0596	.4404	.60	.2257	.2743	1.05	.3531	.1469
.16	.0636	.4364	.61	.2291	.2709	1.06	.3554	.1446
.17	.0675	.4325	.62	.2324	.2676	1.07	.3577	.1423
.18	.0714	.4286	.63	.2357	.2643	1.08	.3599	.1401
.19	.0753	.4247	.64	.2389	.2611	1.09	.3621	.1379
.20	.0793	.4207	.65	.2422	.2578	1.10	.3643	.1357
.21	.0832	.4168	.66	.2454	.2546	1.11	.3665	.1335
.22	.0871	.4129	.67	.2486	.2514	1.12	.3686	.1314
.23	.0910	.4090	.68	.2517	.2483	1.13	.3708	.1292
.24	.0948	.4052	.69	.2549	.2451	1.14	.3729	.1271
.25	.0987	.4013	.70	.2580	.2420	1.15	.3749	.1251
.26	.1026	.3974	.71	.2611	.2389	1.16	.3770	.1230
.27	.1064	.3936	.72	.2642	.2358	1.17	.3790	.1210
.28	.1103	.3897	.73	.2673	.2327	1.18	.3810	.1190
.29	.1141	.3859	.74	.2704	.2296	1.19	.3830	.1170
.30	.1179	.3821	.75	.2734	.2266	1.20	.3849	.1151
.31	.1217	.3783	.76	.2764	.2236	1.21	.2869	.1131
.32	.1255	.3745	.77	.2794	.2206	1.22	.3888	.1112
.33	.1293	.3707	.78	.2823	.2177	1.23	.3907	.1093
.34	.1331	.3669	.79	.2852	.2148	1.24	.3925	.1075
.35	.1368	.3632	.80	.2881	.2119	1.25	.3944	.1056
.36	.1406	.3594	.81	.2910	.2090	1.26	.3962	.1038
.37	.1443	.3557	.82	.2939	.2061	1.27	.3980	.1020
.38	.1480	.3520	.83	.2967	.2033	1.28	.3997	.1003
.39	.1517	.3483	.84	.2995	.2005	1.29	.4015	.0985
.40	.1554	.3446	.85	.3023	.1977	1.30	.4032	.0968
.41	.1591	.3409	.86	.3051	.1949	1.31	.4049	.0951
.42	.1628	.3372	.87	.3078	.1922	1.32	.4066	.0934
.43	.1664	.3336	.88	.3106	.1894	1.33	.4082	.0918
.44	.1700	.3300	.89	.3133	.1867	1.34	.4099	.0901

Column A contains the z scores. Column B gives the proportion of the total curve area between the mean and the z value. Column C contains the proportion of the total area of the curve above the z value.

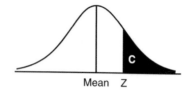

Mean Z

(A) z	(B) Area Between Mean and z	(C) Area Beyond z	(A) z	(B) Area Between Mean and z	(C) Area Beyond z	(A) z	(B) Area Between Mean and z	(C) Area Beyond z
1.35	.4115	.0885	1.80	.4641	.0359	2.25	.4878	.0122
1.36	.4131	.0869	1.81	.4649	.0351	2.26	.4881	.0119
1.37	.4147	.0853	1.82	.4656	.0344	2.27	.4884	.0116
1.38	.4162	.0838	1.83	.4664	.0336	2.28	.4887	.0113
1.39	.4177	.0823	1.84	.4671	.0329	2.29	.4890	.0110
1.40	.4192	.0808	1.85	.4678	.0322	2.30	.4893	.0107
1.41	.4207	.0793	1.86	.4686	.0314	2.31	.4896	.0104
1.42	.4222	.0778	1.87	.4693	.0307	2.32	.4898	.0102
1.43	.4236	.0764	1.88	.4699	.0301	2.33	.4901	.0099
1.44	.4251	.0749	1.89	.4706	.0294	2.34	.4904	.0096
1.45	.4265	.0735	1.90	.4713	.0287	2.35	.4906	.0094
1.46	.4279	.0721	1.91	.4719	.0281	2.36	.4909	.0091
1.47	.4292	.0708	1.92	.4726	.0274	2.37	.4911	.0089
1.48	.4306	.0694	1.93	.4732	.0268	2.38	.4913	.0087
1.49	.4319	.0681	1.94	.4738	.0262	2.39	.4916	.0084
1.50	.4332	.0668	1.95	.4744	.0256	2.40	.4918	.0082
1.51	.4345	.0655	1.96	.4750	.0250	2.41	.4920	.0080
1.52	.4357	.0643	1.97	.4756	.0244	2.42	.4922	.0078
1.53	.4370	.0630	1.98	.4761	.0239	2.43	.4925	.0075
1.54	.4382	.0618	1.99	.4767	.0233	2.44	.4927	.0073
1.55	.4394	.0606	2.00	.4772	.0228	2.45	.4929	.0071
1.56	.4406	.0594	2.01	.4778	.0222	2.46	.4931	.0069
1.57	.4418	.0582	2.02	.4783	.0217	2.47	.4932	.0068
1.58	.4429	.0571	2.03	.4788	.0212	2.48	.4934	.0066
1.59	.4441	.0559	2.04	.4793	.0207	2.49	.4936	.0064
1.60	.4452	.0548	2.05	.4798	.0202	2.50	.4938	.0062
1.61	.4463	.0537	2.06	.4803	.0197	2.51	.4940	.0060
1.62	.4474	.0526	2.07	.4808	.0192	2.52	.4941	.0059
1.63	.4484	.0516	2.08	.4812	.0188	2.53	.4943	.0057
1.64	.4495	.0505	2.09	.4817	.0183	2.54	.4945	.0055
1.65	.4505	.0495	2.10	.4821	.0179	2.55	.4946	.0054
1.66	.4515	.0485	2.11	.4826	.0174	2.56	.4948	.0052
1.67	.4525	.0475	2.12	.4830	.0170	2.57	.4949	.0051
1.68	.4535	.0465	2.13	.4834	.0166	2.58	.4951	.0049
1.69	.4545	.0455	2.14	.4838	.0162	2.59	.4952	.0048
1.70	.4554	.0446	2.15	.4842	.0158	2.60	.4953	.0047
1.71	.4564	.0436	2.16	.4846	.0154	2.61	.4955	.0045
1.72	.4573	.0427	2.17	.4850	.0150	2.62	.4956	.0044
1.73	.4582	.0418	2.18	.4854	.0146	2.63	.4957	.0043
1.74	.4591	.0409	2.19	.4857	.0143	2.64	.4959	.0041
1.75	.4599	.0401	2.20	.4861	.0139	2.65	.4960	.0040
1.76	.4608	.0392	2.21	.4864	.0136	2.66	.4961	.0039
1.77	.4616	.0384	2.22	.4868	.0132	2.67	.4962	.0038
1.78	.4625	.0375	2.23	.4871	.0129	2.68	.4963	.0037
1.79	.4633	.0367	2.24	.4875	.0125	2.69	.4964	.0036

For a negative z, Column B contains the proportion of the area under the total curve between the negative z and the mean. Column C contains the proportion of the area under the curve less than the negative z value.

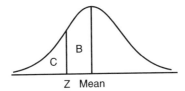

(A)	(B) Area Between	(C) Area Beyond	(A)	(B) Area Between	(C) Area Beyond	(A)	(B) Area Between	(C) Area Beyond
z	Mean and z	z	z	Mean and z	z	z	Mean and z	z
2.70	.4965	.0035	2.95	.4984	.0016	3.20	.4993	.0007
2.71	.4966	.0034	2.96	.4985	.0015	3.21	.4993	.0007
2.72	.4967	.0033	2.97	.4985	.0015	3.22	.4994	.0006
2.73	.4968	.0032	2.98	.4986	.0014	3.23	.4994	.0006
2.74	.4969	.0031	2.99	.4986	.0014	3.24	.4994	.0006
2.75	.4970	.0030	3.00	.4987	.0013	3.30	.4995	.0005
2.76	.4971	.0029	3.01	.4987	.0013	3.40	.4997	.0003
2.77	.4972	.0028	3.02	.4987	.0013	3.50	.4998	.0002
2.78	.4973	.0027	3.03	.4988	.0012	3.60	.4998	.0002
2.79	.4974	.0026	3.04	.4988	.0012	3.70	.4999	.0001
2.80	.4974	.0026	3.05	.4989	.0011	3.80	.49993	.00007
2.81	.4975	.0025	3.06	.4989	.0011	3.90	.49995	.00005
2.82	.4976	.0024	3.07	.4989	.0011	4.00	.49997	.00003
2.83	.4977	.0023	3.08	.4990	.0010			
2.84	.4977	.0023	3.09	.4990	.0010			
2.85	.4978	.0022	3.10	.4990	.0010			
2.86	.4979	.0021	3.11	.4991	.0009			
2.87	.4979	.0021	3.12	.4991	.0009			
2.88	.4980	.0020	3.13	.4991	.0009			
2.89	.4981	.0019	3.14	.4992	.0008			
2.90	.4981	.0019	3.15	.4992	.0008			
2.91	.4982	.0018	3.16	.4992	.0008			
2.92	.4982	.0018	3.17	.4992	.0008			
2.93	.4983	.0017	3.18	.4993	.0007			
2.94	.4984	.0016	3.19	.4993	.0007			

Appendix B
The *t* Distribution

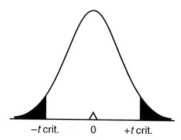

−*t* crit.   0   +*t* crit.

Two-tailed or Nondirectional Test

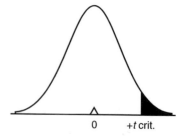

0   +*t* crit.

One-tailed or Directional Test

LEVEL OF SIGNIFICANCE

df	.05	.01	.001
1	12.706	63.657	636.62
2	4.303	9.925	31.598
3	3.182	5.841	12.924
4	2.776	4.604	8.610
5	2.571	4.032	6.869
6	2.447	3.707	5.959
7	2.365	3.499	5.408
8	2.306	3.355	5.041
9	2.262	3.250	4.781
10	2.228	3.169	4.587
11	2.201	3.106	4.437
12	2.179	3.055	4.318
13	2.160	3.012	4.221
14	2.145	2.977	4.140
15	2.131	2.947	4.073
16	2.120	2.921	4.015
17	2.110	2.898	3.965
18	2.101	2.878	3.922
19	2.093	2.861	3.883
20	2.086	2.845	3.850
21	2.080	2.831	3.819
22	2.074	2.819	3.792
23	2.069	2.807	3.767
24	2.064	2.797	3.745
25	2.060	2.787	3.725
26	2.056	2.779	3.707
27	2.052	2.771	3.690
28	2.048	2.763	3.674
29	2.045	2.756	3.659
30	2.042	2.750	3.646
40	2.021	2.704	3.551
60	2.000	2.660	3.460
120	1.980	2.617	3.373
∞	1.960	2.576	3.291

LEVEL OF SIGNIFICANCE

df	.05	.01	.001
1	6.314	31.821	318.31
2	2.920	6.965	22.326
3	2.353	4.541	10.213
4	2.132	3.747	7.173
5	2.015	3.365	5.893
6	1.943	3.143	5.308
7	1.895	2.998	4.785
8	1.860	2.896	4.501
9	1.833	2.821	4.297
10	1.812	2.764	4.144
11	1.796	2.718	4.025
12	1.782	2.681	3.930
13	1.771	2.650	3.852
14	1.761	2.624	3.787
15	1.753	2.602	3.733
16	1.746	2.583	3.686
17	1.740	2.567	3.646
18	1.734	2.552	3.610
19	1.729	2.539	3.579
20	1.725	2.528	3.552
21	1.721	2.518	3.527
22	1.717	2.508	3.505
23	1.714	2.500	3.485
24	1.711	2.492	3.467
25	1.708	2.485	3.450
26	1.706	2.479	3.435
27	1.703	2.473	3.421
28	1.701	2.467	3.408
29	1.699	2.462	3.396
30	1.697	2.457	3.385
40	1.684	2.423	3.307
60	1.671	2.390	3.232
120	1.658	2.358	3.160
∞	1.645	2.326	3.090

Appendix C
Spearman's Correlation

Critical Values for the Spearman Rank-Order Correlation Coefficient

Number of Pairs n	Level of significance for a one-tailed test			
	.05	.025	.01	.005
	Level of significance for a two-tailed test			
	.10	.05	.02	.01
5	0.900	1.000	1.000	—
6	0.829	0.886	0.943	1.000
7	0.714	0.786	0.893	0.929
8	0.643	0.738	0.833	0.881
9	0.600	0.683	0.783	0.833
10	0.564	0.648	0.746	0.794
12	0.506	0.591	0.712	0.777
14	0.456	0.544	0.645	0.715
16	0.425	0.506	0.601	0.665
18	0.399	0.475	0.564	0.625
20	0.377	0.450	0.534	0.591
22	0.359	0.428	0.508	0.562
24	0.343	0.409	0.485	0.537
26	0.329	0.392	0.465	0.515
28	0.317	0.377	0.448	0.496
30	0.306	0.364	0.432	0.478

Source: Glasser, G. J. and R. F. Winter, 'Critical Values of the Coefficient of Rank Correlation for Testing the Hypothesis of Independence.' *Biometrika*, 48, 444 (1961).
Notes: Reject null hypothesis of obtained rho is equal to or greater than tabled value.

Appendix D
The CHI Square Distribution

Critical Values of CHI Square

df	α levels				
	.10	.05	.02	.01	.001
1	2.71	3.84	5.41	6.64	10.38
2	4.60	5.99	7.82	9.21	13.82
3	6.25	7.82	9.84	11.34	16.27
4	7.78	9.49	11.67	13.28	18.46
5	9.24	11.07	13.39	15.09	20.52
6	10.64	12.59	15.03	16.81	22.46
7	12.02	14.07	16.62	18.48	24.32
8	13.36	15.51	18.17	20.09	26.12
9	14.68	16.92	19.68	21.67	27.88
10	15.99	18.31	21.16	23.21	29.59
11	17.28	19.68	22.62	24.72	31.26
12	18.55	21.03	24.05	26.22	32.91
13	19.81	22.36	25.47	27.69	34.53
14	21.06	23.68	26.87	29.14	36.12
15	22.31	25.00	28.26	30.58	37.70
16	23.54	26.30	29.63	32.00	39.25
17	24.77	27.59	31.00	33.41	40.79
18	25.99	28.87	32.35	34.80	42.31
19	27.20	30.14	33.69	36.19	43.82
20	28.41	31.41	35.02	37.57	45.32
21	29.62	32.67	36.34	38.93	46.80
22	30.81	33.92	37.66	40.29	48.27
23	32.01	35.17	38.97	41.64	49.73
24	33.20	36.42	40.27	42.98	51.18
25	34.38	37.65	41.57	44.31	52.62
26	35.56	38.88	42.86	45.64	54.05
27	36.74	40.11	44.14	46.96	55.48
28	37.92	41.34	45.42	48.28	56.89
29	39.09	42.56	46.69	49.59	58.30
30	40.26	43.77	47.96	50.89	59.70

Source: This table is taken from Table IV of Fisher and Yates. *Statistical Tables for Biological, Agricultural and Medical Research*, published by Longman Group Ltd., London (previously published by Oliver and Boyd, Ltd., Edinburgh), and by permission of the authors and publishers.
Note: Reject null hypothesis if obtained chi square is equal to or greater than the tabled value.

Appendix E
The F Distribution

.05 Level of significance (upper numbers)
.01 Level of significance (lower numbers)

Critical Values of F

.05 Level (on top) and .01 Level (below) α Levels for the distribution of F

df (denom)		Degrees of freedom (for the numerator)																							
		1	2	3	4	5	6	7	8	9	10	11	12	14	16	20	24	30	40	50	75	100	200	500	∞
1	.05	161	200	216	225	230	234	237	239	241	242	243	244	245	246	248	249	250	251	252	253	253	254	254	254
	.01	4,052	4,999	5,403	5,625	5,764	5,859	5,928	5,981	6,022	6,056	6,082	6,106	6,142	6,169	6,208	6,234	6,258	6,286	6,302	6,323	6,334	6,352	6,361	6,366
2	.05	18.51	19.00	19.16	19.25	19.30	19.33	19.36	19.37	19.38	19.39	19.40	19.41	19.42	19.43	19.44	19.45	19.46	19.47	19.47	19.48	19.49	19.49	19.50	19.50
	.01	98.49	99.00	99.17	99.25	99.30	99.33	99.34	99.36	99.38	99.40	99.41	99.42	99.43	99.44	99.45	99.46	99.47	99.48	99.48	99.49	99.49	99.49	99.50	99.50
3	.05	10.13	9.55	9.28	9.12	9.01	8.94	8.88	8.84	8.81	8.78	8.76	8.74	8.71	8.69	8.66	8.64	8.62	8.60	8.58	8.57	8.56	8.54	8.54	8.53
	.01	34.12	30.82	29.46	28.71	28.24	27.91	27.67	27.49	27.34	27.23	27.13	27.05	26.92	26.83	26.69	26.60	26.50	26.41	26.35	26.27	26.23	26.18	26.14	26.12
4	.05	7.71	6.94	6.59	6.39	6.26	6.16	6.09	6.04	6.00	5.96	5.93	5.91	5.87	5.84	5.80	5.77	5.74	5.71	5.70	5.68	5.66	5.65	5.64	5.63
	.01	21.20	18.00	16.69	15.98	15.52	15.21	14.98	14.80	14.66	14.54	14.45	14.37	14.24	14.15	14.02	13.93	13.83	13.74	13.69	13.61	13.57	13.52	13.48	13.46
5	.05	6.61	5.79	5.41	5.19	5.05	4.95	4.88	4.82	4.78	4.74	4.70	4.68	4.64	4.60	4.56	4.53	4.50	4.46	4.44	4.42	4.40	4.38	4.37	4.36
	.01	16.26	13.27	12.06	11.39	10.97	10.67	10.45	10.27	10.15	10.05	9.96	9.89	9.77	9.68	9.55	9.47	9.38	9.29	9.24	9.17	9.13	9.07	9.04	9.02
6	.05	5.99	5.14	4.76	4.53	4.39	4.28	4.21	4.15	4.10	4.06	4.03	4.00	3.96	3.92	3.87	3.84	3.81	3.77	3.75	3.72	3.71	3.69	3.68	3.67
	.01	13.74	10.92	9.78	9.15	8.75	8.47	8.26	8.10	7.98	7.87	7.79	7.72	7.60	7.52	7.39	7.31	7.23	7.14	7.09	7.02	6.99	6.94	6.90	6.88
7	.05	5.59	4.74	4.35	4.12	3.97	3.87	3.79	3.73	3.68	3.63	3.60	3.57	3.52	3.49	3.44	3.41	3.38	3.34	3.32	3.29	3.28	3.25	3.24	3.23
	.01	12.25	9.55	8.45	7.85	7.46	7.19	7.00	6.84	6.71	6.62	6.54	6.47	6.35	6.27	6.15	6.07	5.98	5.90	5.85	5.78	5.75	5.70	5.67	5.65
8	.05	5.32	4.46	4.07	3.84	3.69	3.58	3.50	3.44	3.39	3.34	3.31	3.28	3.23	3.20	3.15	3.12	3.08	3.05	3.03	3.00	2.98	2.96	2.94	2.93
	.01	11.26	8.65	7.59	7.01	6.63	6.37	6.19	6.03	5.91	5.82	5.74	5.67	5.56	5.48	5.36	5.28	5.20	5.11	5.06	5.00	4.96	4.91	4.88	4.86
9	.05	5.12	4.26	3.86	3.63	3.48	3.37	3.29	3.23	3.18	3.13	3.10	3.07	3.02	2.98	2.93	2.90	2.86	2.82	2.80	2.77	2.76	2.73	2.72	2.71
	.01	10.56	8.02	6.99	6.42	6.06	5.80	5.62	5.47	5.35	5.26	5.18	5.11	5.00	4.92	4.80	4.73	4.64	4.56	4.51	4.45	4.41	4.36	4.33	4.31
10	.05	4.96	4.10	3.71	3.48	3.33	3.22	3.14	3.07	3.02	2.97	2.94	2.91	2.86	2.82	2.77	2.74	2.70	2.67	2.64	2.61	2.59	2.56	2.55	2.54
	.01	10.04	7.56	6.55	5.99	5.64	5.39	5.21	5.06	4.95	4.85	4.78	4.71	4.60	4.52	4.41	4.33	4.25	4.17	4.12	4.05	4.01	3.96	3.93	3.91

Degrees of freedom (for the denominator)

Degrees of freedom in numerator

df (denom)	1	2	3	4	5	6	7	8	9	10	11	12	14	16	20	24	30	40	50	75	100	200	500	∞
11	4.84 / 9.65	3.98 / 7.20	3.59 / 6.22	3.36 / 5.67	3.20 / 5.32	3.09 / 5.07	3.01 / 4.88	2.95 / 4.74	2.90 / 4.63	2.86 / 4.54	2.82 / 4.46	2.79 / 4.40	2.74 / 4.29	2.70 / 4.21	2.65 / 4.10	2.61 / 4.02	2.57 / 3.94	2.53 / 3.86	2.50 / 3.80	2.47 / 3.74	2.45 / 3.70	2.42 / 3.66	2.41 / 3.62	2.40 / 3.60
12	4.75 / 9.33	3.88 / 6.93	3.49 / 5.95	3.26 / 5.41	3.11 / 5.06	3.00 / 4.82	2.92 / 4.65	2.85 / 4.50	2.80 / 4.39	2.76 / 4.30	2.72 / 4.22	2.69 / 4.16	2.64 / 4.05	2.60 / 3.98	2.54 / 3.86	2.50 / 3.78	2.46 / 3.70	2.42 / 3.61	2.40 / 3.56	2.36 / 3.49	2.35 / 3.46	2.32 / 3.41	2.31 / 3.38	2.30 / 3.36
13	4.67 / 9.07	3.80 / 6.70	3.41 / 5.74	3.18 / 5.20	3.02 / 4.86	2.92 / 4.62	2.84 / 4.44	2.77 / 4.30	2.72 / 4.19	2.67 / 4.10	2.63 / 4.02	2.60 / 3.96	2.55 / 3.85	2.51 / 3.78	2.46 / 3.67	2.42 / 3.59	2.38 / 3.51	2.34 / 3.42	2.32 / 3.37	2.28 / 3.30	2.26 / 3.27	2.24 / 3.21	2.22 / 3.18	2.21 / 3.16
14	4.60 / 8.86	3.74 / 6.51	3.34 / 5.56	3.11 / 5.03	2.96 / 4.69	2.85 / 4.46	2.77 / 4.28	2.70 / 4.14	2.65 / 4.03	2.60 / 3.94	2.56 / 3.86	2.53 / 3.80	2.48 / 3.70	2.44 / 3.62	2.39 / 3.51	2.35 / 3.43	2.31 / 3.34	2.27 / 3.26	2.24 / 3.21	2.21 / 3.14	2.19 / 3.11	2.16 / 3.06	2.14 / 3.02	2.13 / 3.00
15	4.54 / 8.68	3.68 / 6.36	3.29 / 5.42	3.06 / 4.89	2.90 / 4.56	2.79 / 4.32	2.70 / 4.14	2.64 / 4.00	2.59 / 3.89	2.55 / 3.80	2.51 / 3.73	2.48 / 3.67	2.43 / 3.56	2.39 / 3.48	2.33 / 3.36	2.29 / 3.29	2.25 / 3.20	2.21 / 3.12	2.18 / 3.07	2.15 / 3.00	2.12 / 2.97	2.10 / 2.92	2.08 / 2.89	2.07 / 2.87
16	4.49 / 8.53	3.63 / 6.23	3.24 / 5.29	3.01 / 4.77	2.85 / 4.44	2.74 / 4.20	2.66 / 4.03	2.59 / 3.89	2.54 / 3.78	2.49 / 3.69	2.45 / 3.61	2.42 / 3.55	2.37 / 3.45	2.33 / 3.37	2.28 / 3.25	2.24 / 3.18	2.20 / 3.10	2.16 / 3.01	2.13 / 2.96	2.09 / 2.89	2.07 / 2.86	2.04 / 2.80	2.02 / 2.77	2.01 / 2.75
17	4.45 / 8.40	3.59 / 6.11	3.20 / 5.18	2.96 / 4.67	2.81 / 4.34	2.70 / 4.10	2.62 / 3.93	2.55 / 3.79	2.50 / 3.68	2.45 / 3.59	2.41 / 3.52	2.38 / 3.45	2.33 / 3.35	2.29 / 3.27	2.23 / 3.16	2.19 / 3.08	2.15 / 3.00	2.11 / 2.92	2.08 / 2.86	2.04 / 2.79	2.02 / 2.76	1.99 / 2.70	1.97 / 2.67	1.96 / 2.65
18	4.41 / 8.28	3.55 / 6.01	3.16 / 5.08	2.93 / 4.58	2.77 / 4.25	2.66 / 4.01	2.58 / 3.85	2.51 / 3.71	2.46 / 3.60	2.41 / 3.51	2.37 / 3.44	2.34 / 3.37	2.29 / 3.27	2.25 / 3.19	2.19 / 3.07	2.15 / 3.00	2.11 / 2.91	2.07 / 2.83	2.04 / 2.78	2.00 / 2.71	1.98 / 2.68	1.95 / 2.62	1.93 / 2.58	1.92 / 2.57
19	4.38 / 8.18	3.52 / 5.93	3.13 / 5.01	2.90 / 4.50	2.74 / 4.17	2.63 / 3.94	2.55 / 3.77	2.48 / 3.63	2.43 / 3.52	2.38 / 3.43	2.34 / 3.36	2.31 / 3.30	2.26 / 3.19	2.21 / 3.12	2.15 / 3.00	2.11 / 2.92	2.07 / 2.84	2.02 / 2.76	2.00 / 2.70	1.96 / 2.63	1.94 / 2.60	1.91 / 2.54	1.90 / 2.51	1.88 / 2.49
20	4.35 / 8.10	3.49 / 5.85	3.10 / 4.94	2.87 / 4.43	2.71 / 4.10	2.60 / 3.87	2.52 / 3.71	2.45 / 3.56	2.40 / 3.45	2.35 / 3.37	2.31 / 3.30	2.28 / 3.23	2.23 / 3.13	2.18 / 3.05	2.12 / 2.94	2.08 / 2.86	2.04 / 2.77	1.99 / 2.69	1.96 / 2.63	1.92 / 2.56	1.90 / 2.53	1.87 / 2.47	1.85 / 2.44	1.84 / 2.42
21	4.32 / 8.02	3.47 / 5.78	3.07 / 4.87	2.84 / 4.37	2.68 / 4.04	2.57 / 3.81	2.49 / 3.65	2.42 / 3.51	2.37 / 3.40	2.32 / 3.31	2.28 / 3.24	2.25 / 3.17	2.20 / 3.07	2.15 / 2.99	2.09 / 2.88	2.05 / 2.80	2.00 / 2.72	1.96 / 2.63	1.93 / 2.58	1.89 / 2.51	1.87 / 2.47	1.84 / 2.42	1.82 / 2.38	1.81 / 2.36
22	4.30 / 7.94	3.44 / 5.72	3.05 / 4.82	2.82 / 4.31	2.66 / 3.99	2.55 / 3.76	2.47 / 3.59	2.40 / 3.45	2.35 / 3.35	2.30 / 3.26	2.26 / 3.18	2.23 / 3.12	2.18 / 3.02	2.13 / 2.94	2.07 / 2.83	2.03 / 2.75	1.98 / 2.67	1.93 / 2.58	1.91 / 2.53	1.87 / 2.46	1.84 / 2.42	1.81 / 2.37	1.80 / 2.33	1.78 / 2.31
23	4.28 / 7.88	3.42 / 5.66	3.03 / 4.76	2.80 / 4.26	2.64 / 3.94	2.53 / 3.71	2.45 / 3.54	2.38 / 3.41	2.32 / 3.30	2.28 / 3.21	2.24 / 3.14	2.20 / 3.07	2.14 / 2.97	2.10 / 2.89	2.04 / 2.78	2.00 / 2.70	1.96 / 2.62	1.91 / 2.53	1.88 / 2.48	1.84 / 2.41	1.82 / 2.37	1.79 / 2.32	1.77 / 2.28	1.76 / 2.26
27	4.21 / 7.68	3.35 / 5.49	2.96 / 4.60	2.73 / 4.11	2.57 / 3.79	2.46 / 3.56	2.37 / 3.39	2.30 / 3.26	2.25 / 3.14	2.20 / 3.06	2.16 / 2.98	2.13 / 2.93	2.08 / 2.83	2.03 / 2.74	1.97 / 2.63	1.93 / 2.55	1.88 / 2.47	1.84 / 2.38	1.80 / 2.33	1.76 / 2.25	1.74 / 2.21	1.71 / 2.16	1.68 / 2.12	1.67 / 2.10
28	4.20 / 7.64	3.34 / 5.45	2.95 / 4.57	2.71 / 4.07	2.56 / 3.76	2.44 / 3.53	2.36 / 3.36	2.29 / 3.23	2.24 / 3.11	2.19 / 3.03	2.15 / 2.95	2.12 / 2.90	2.06 / 2.80	2.02 / 2.71	1.96 / 2.60	1.91 / 2.52	1.87 / 2.44	1.81 / 2.35	1.78 / 2.30	1.75 / 2.22	1.72 / 2.18	1.69 / 2.13	1.67 / 2.09	1.65 / 2.06
29	4.18 / 7.60	3.33 / 5.42	2.93 / 4.54	2.70 / 4.04	2.54 / 3.73	2.43 / 3.50	2.35 / 3.33	2.28 / 3.20	2.22 / 3.08	2.18 / 3.00	2.14 / 2.92	2.10 / 2.87	2.05 / 2.77	2.00 / 2.68	1.94 / 2.57	1.90 / 2.49	1.85 / 2.41	1.80 / 2.32	1.77 / 2.27	1.73 / 2.19	1.71 / 2.15	1.68 / 2.10	1.65 / 2.06	1.64 / 2.03

Degrees of freedom (for the denominator)

Degrees of freedom in numerator

df (denom)	1	2	3	4	5	6	7	8	9	10	11	12	14	16	20	24	30	40	50	75	100	200	500	∞
30	4.17 / 7.56	3.32 / 5.39	2.92 / 4.51	2.69 / 4.02	2.53 / 3.70	2.42 / 3.47	2.34 / 3.30	2.27 / 3.17	2.21 / 3.06	2.16 / 2.98	2.12 / 2.90	2.09 / 2.84	2.04 / 2.74	1.99 / 2.66	1.93 / 2.55	1.89 / 2.47	1.84 / 2.38	1.79 / 2.29	1.76 / 2.24	1.72 / 2.16	1.69 / 2.13	1.66 / 2.07	1.64 / 2.03	1.62 / 2.01
32	4.15 / 7.50	3.30 / 5.34	2.90 / 4.46	2.67 / 3.97	2.51 / 3.66	2.40 / 3.42	2.32 / 3.25	2.25 / 3.12	2.19 / 3.01	2.14 / 2.94	2.10 / 2.86	2.07 / 2.80	2.02 / 2.70	1.97 / 2.62	1.91 / 2.51	1.86 / 2.42	1.82 / 2.34	1.76 / 2.25	1.74 / 2.20	1.69 / 2.12	1.67 / 2.08	1.64 / 2.02	1.61 / 1.98	1.59 / 1.96
34	4.13 / 7.44	3.28 / 5.29	2.88 / 4.42	2.65 / 3.93	2.49 / 3.61	2.38 / 3.38	2.30 / 3.21	2.23 / 3.08	2.17 / 2.97	2.12 / 2.89	2.08 / 2.82	2.05 / 2.76	2.00 / 2.66	1.95 / 2.58	1.89 / 2.47	1.84 / 2.38	1.80 / 2.30	1.74 / 2.21	1.71 / 2.15	1.67 / 2.08	1.64 / 2.04	1.61 / 1.98	1.59 / 1.94	1.57 / 1.91
36	4.11 / 7.39	3.26 / 5.25	2.86 / 4.38	2.63 / 3.89	2.48 / 3.58	2.36 / 3.35	2.28 / 3.18	2.21 / 3.04	2.15 / 2.94	2.10 / 2.86	2.06 / 2.78	2.03 / 2.72	1.98 / 2.62	1.93 / 2.54	1.87 / 2.43	1.82 / 2.35	1.78 / 2.26	1.72 / 2.17	1.69 / 2.12	1.65 / 2.04	1.62 / 2.00	1.59 / 1.94	1.56 / 1.90	1.55 / 1.87
38	4.10 / 7.35	3.25 / 5.21	2.85 / 4.34	2.62 / 3.86	2.46 / 3.54	2.35 / 3.32	2.26 / 3.15	2.19 / 3.02	2.14 / 2.91	2.09 / 2.82	2.05 / 2.75	2.02 / 2.69	1.96 / 2.59	1.92 / 2.51	1.85 / 2.40	1.80 / 2.32	1.76 / 2.22	1.71 / 2.14	1.67 / 2.08	1.63 / 2.00	1.60 / 1.97	1.57 / 1.90	1.54 / 1.86	1.53 / 1.84
40	4.08 / 7.31	3.23 / 5.18	2.84 / 4.31	2.61 / 3.83	2.45 / 3.51	2.34 / 3.29	2.25 / 3.12	2.18 / 2.99	2.12 / 2.88	2.07 / 2.80	2.04 / 2.73	2.00 / 2.66	1.95 / 2.56	1.90 / 2.49	1.84 / 2.37	1.79 / 2.29	1.74 / 2.20	1.69 / 2.11	1.66 / 2.05	1.61 / 1.97	1.59 / 1.94	1.55 / 1.88	1.53 / 1.84	1.51 / 1.81
42	4.07 / 7.27	3.22 / 5.15	2.83 / 4.29	2.59 / 3.80	2.44 / 3.49	2.32 / 3.26	2.24 / 3.10	2.17 / 2.96	2.11 / 2.86	2.06 / 2.77	2.02 / 2.70	1.99 / 2.64	1.94 / 2.54	1.89 / 2.46	1.82 / 2.35	1.78 / 2.26	1.73 / 2.17	1.68 / 2.08	1.64 / 2.02	1.60 / 1.94	1.57 / 1.91	1.54 / 1.85	1.51 / 1.80	1.49 / 1.78
44	4.06 / 7.24	3.21 / 5.12	2.82 / 4.26	2.58 / 3.78	2.43 / 3.46	2.31 / 3.24	2.23 / 3.07	2.16 / 2.94	2.10 / 2.84	2.05 / 2.75	2.01 / 2.68	1.98 / 2.62	1.92 / 2.52	1.88 / 2.44	1.81 / 2.32	1.76 / 2.24	1.72 / 2.15	1.66 / 2.06	1.63 / 2.00	1.58 / 1.92	1.56 / 1.88	1.52 / 1.82	1.50 / 1.78	1.48 / 1.75
46	4.05 / 7.21	3.20 / 5.10	2.81 / 4.24	2.57 / 3.76	2.42 / 3.44	2.30 / 3.22	2.22 / 3.05	2.14 / 2.92	2.09 / 2.82	2.04 / 2.73	2.00 / 2.66	1.97 / 2.60	1.91 / 2.50	1.87 / 2.42	1.80 / 2.30	1.75 / 2.22	1.71 / 2.13	1.65 / 2.04	1.62 / 1.98	1.57 / 1.90	1.54 / 1.86	1.51 / 1.80	1.48 / 1.76	1.46 / 1.72
48	4.04 / 7.19	3.19 / 5.08	2.80 / 4.22	2.56 / 3.74	2.41 / 3.42	2.30 / 3.20	2.21 / 3.04	2.14 / 2.90	2.08 / 2.80	2.03 / 2.71	1.99 / 2.64	1.96 / 2.58	1.90 / 2.48	1.86 / 2.40	1.79 / 2.28	1.74 / 2.20	1.70 / 2.11	1.64 / 2.02	1.61 / 1.96	1.56 / 1.88	1.53 / 1.84	1.50 / 1.78	1.47 / 1.73	1.45 / 1.70
50	4.03 / 7.17	3.18 / 5.06	2.79 / 4.20	2.56 / 3.72	2.40 / 3.41	2.29 / 3.18	2.20 / 3.02	2.13 / 2.88	2.07 / 2.78	2.02 / 2.70	1.98 / 2.62	1.95 / 2.56	1.90 / 2.46	1.85 / 2.39	1.78 / 2.26	1.74 / 2.18	1.69 / 2.10	1.63 / 2.00	1.60 / 1.94	1.55 / 1.86	1.52 / 1.82	1.48 / 1.76	1.46 / 1.71	1.44 / 1.68
55	4.02 / 7.12	3.17 / 5.01	2.78 / 4.16	2.54 / 3.68	2.38 / 3.37	2.27 / 3.15	2.18 / 2.98	2.11 / 2.85	2.05 / 2.75	2.00 / 2.66	1.97 / 2.59	1.93 / 2.53	1.88 / 2.43	1.83 / 2.35	1.76 / 2.23	1.72 / 2.15	1.67 / 2.06	1.61 / 1.96	1.58 / 1.90	1.52 / 1.82	1.50 / 1.78	1.46 / 1.71	1.43 / 1.66	1.41 / 1.64
60	4.00 / 7.08	3.15 / 4.98	2.76 / 4.13	2.52 / 3.65	2.37 / 3.34	2.25 / 3.12	2.17 / 2.95	2.10 / 2.82	2.04 / 2.72	1.99 / 2.63	1.95 / 2.56	1.92 / 2.50	1.86 / 2.40	1.81 / 2.32	1.75 / 2.20	1.70 / 2.12	1.65 / 2.03	1.59 / 1.93	1.56 / 1.87	1.50 / 1.79	1.48 / 1.74	1.44 / 1.68	1.41 / 1.63	1.39 / 1.60
65	3.99 / 7.04	3.14 / 4.95	2.75 / 4.10	2.51 / 3.62	2.36 / 3.31	2.24 / 3.09	2.15 / 2.93	2.08 / 2.79	2.02 / 2.70	1.98 / 2.61	1.94 / 2.54	1.90 / 2.47	1.85 / 2.37	1.80 / 2.30	1.73 / 2.18	1.68 / 2.09	1.63 / 2.00	1.57 / 1.90	1.54 / 1.84	1.49 / 1.76	1.46 / 1.71	1.42 / 1.64	1.39 / 1.60	1.37 / 1.56
70	3.98 / 7.01	3.13 / 4.92	2.74 / 4.08	2.50 / 3.60	2.35 / 3.29	2.23 / 3.07	2.14 / 2.91	2.07 / 2.77	2.01 / 2.67	1.97 / 2.59	1.93 / 2.51	1.89 / 2.45	1.84 / 2.35	1.79 / 2.28	1.72 / 2.15	1.67 / 2.07	1.62 / 1.98	1.56 / 1.88	1.53 / 1.82	1.47 / 1.74	1.45 / 1.69	1.40 / 1.62	1.37 / 1.56	1.35 / 1.53
80	3.96 / 6.96	3.11 / 4.88	2.72 / 4.04	2.48 / 3.56	2.33 / 3.25	2.21 / 3.04	2.12 / 2.87	2.05 / 2.74	1.99 / 2.64	1.95 / 2.55	1.91 / 2.48	1.88 / 2.41	1.82 / 2.32	1.77 / 2.24	1.70 / 2.11	1.65 / 2.03	1.60 / 1.94	1.54 / 1.84	1.51 / 1.78	1.45 / 1.70	1.42 / 1.65	1.38 / 1.57	1.35 / 1.52	1.32 / 1.49

Degrees of freedom (for the denominator)

Degrees of freedom in numerator

Degrees of freedom (for the denominator)	1	2	3	4	5	6	7	8	9	10	11	12	14	16	20	24	30	40	50	75	100	200	500	∞
100	3.94 6.90	3.09 4.82	2.70 3.98	2.46 3.51	2.30 3.20	2.19 2.99	2.10 2.82	2.03 2.69	1.97 2.59	1.92 2.51	1.88 2.43	1.85 2.36	1.79 2.26	1.75 2.19	1.68 2.06	1.63 1.98	1.57 1.89	1.51 1.79	1.48 1.73	1.42 1.64	1.39 1.59	1.34 1.51	1.30 1.46	1.28 1.43
125	3.92 6.84	3.07 4.78	2.68 3.94	2.44 3.47	2.29 3.17	2.17 2.95	2.08 2.79	2.01 2.65	1.95 2.56	1.90 2.47	1.86 2.40	1.83 2.33	1.77 2.23	1.72 2.15	1.65 2.03	1.60 1.94	1.55 1.85	1.49 1.75	1.45 1.68	1.39 1.59	1.36 1.54	1.31 1.46	1.27 1.40	1.25 1.37
150	3.91 6.81	3.06 4.75	2.67 3.91	2.43 3.44	2.27 3.14	2.16 2.92	2.07 2.76	2.00 2.62	1.94 2.53	1.89 2.44	1.85 2.37	1.82 2.30	1.76 2.20	1.71 2.12	1.64 2.00	1.59 1.91	1.54 1.83	1.47 1.72	1.44 1.66	1.37 1.56	1.34 1.51	1.29 1.43	1.25 1.37	1.22 1.33
200	3.89 6.76	3.04 4.71	2.65 3.88	2.41 3.41	2.26 3.11	2.14 2.90	2.05 2.73	1.98 2.60	1.92 2.50	1.87 2.41	1.83 2.34	1.80 2.28	1.74 2.17	1.69 2.09	1.62 1.97	1.57 1.88	1.52 1.79	1.45 1.69	1.42 1.62	1.35 1.53	1.32 1.48	1.26 1.39	1.22 1.33	1.19 1.28
400	3.86 6.70	3.02 4.66	2.62 3.83	2.39 3.36	2.23 3.06	2.12 2.85	2.03 2.69	1.96 2.55	1.90 2.46	1.85 2.37	1.81 2.29	1.78 2.23	1.72 2.12	1.67 2.04	1.60 1.92	1.54 1.84	1.49 1.74	1.42 1.64	1.38 1.57	1.32 1.47	1.28 1.42	1.22 1.32	1.16 1.24	1.13 1.19
1000	3.85 6.66	3.00 4.62	2.61 3.80	2.38 3.34	2.22 3.04	2.10 2.82	2.02 2.66	1.95 2.53	1.89 2.43	1.84 2.34	1.80 2.26	1.76 2.20	1.70 2.09	1.65 2.01	1.58 1.89	1.53 1.81	1.47 1.71	1.41 1.61	1.36 1.54	1.30 1.44	1.26 1.38	1.19 1.28	1.13 1.19	1.08 1.11
∞	3.84 6.64	2.99 4.60	2.60 3.78	2.37 3.32	2.21 3.02	2.09 2.80	2.01 2.64	1.94 2.51	1.88 2.41	1.83 2.32	1.79 2.24	1.75 2.18	1.69 2.07	1.64 1.99	1.57 1.87	1.52 1.79	1.46 1.69	1.40 1.59	1.35 1.52	1.28 1.41	1.24 1.36	1.17 1.25	1.11 1.15	1.00 1.00

Source: Reproduced by permission from *Statistical Methods*, 5th edition by George B. Snedecor, copyright 1956 by the Iowa State University Press.
Notes: Reject null hypothesis if F ratio is equal to or greater than tabled value.

Appendix F
Tukey's Table

q Values

Percentage Points of the Studentized Range

r = number of means

Error df	A	2	3	4	5	6	7	8	9	10	11
5	.05	3.64	4.60	5.22	5.67	6.03	6.33	6.58	6.80	6.99	7.17
	.01	5.70	6.98	7.80	8.42	8.91	9.32	9.67	9.97	10.24	10.48
6	.05	3.46	4.34	4.90	5.30	5.63	5.94	6.12	6.32	6.49	6.65
	.01	5.24	6.33	7.03	7.56	7.97	8.32	8.61	8.87	9.10	9.40
7	.05	3.34	4.16	4.68	5.06	5.36	5.61	5.82	6.00	6.16	6.30
	.01	4.95	5.92	6.54	7.01	7.37	7.68	7.94	8.17	8.37	8.55
8	.05	3.26	4.04	4.53	4.89	5.17	5.40	5.60	5.77	5.92	6.05
	.01	4.75	5.64	6.20	6.62	6.96	7.24	7.47	7.68	7.86	8.03
9	.05	3.20	3.95	4.41	4.76	5.02	5.24	5.43	5.59	5.74	5.87
	.01	4.60	5.43	5.96	6.35	6.66	6.91	7.13	7.33	7.49	7.65
10	.05	3.15	3.88	4.33	4.65	4.91	5.12	5.30	5.46	5.60	5.72
	.01	4.48	5.27	5.77	6.14	6.43	6.67	6.87	7.05	7.21	7.36
11	.05	3.11	3.82	4.26	4.57	4.82	5.03	5.20	5.35	5.49	5.61
	.01	4.39	5.15	5.62	5.97	6.25	6.48	6.67	6.84	6.99	7.13
12	.05	3.08	3.77	4.20	4.51	4.75	4.95	5.12	5.27	5.39	5.51
	.01	4.32	5.05	5.50	5.84	6.10	6.32	6.51	6.67	6.81	6.94
13	.05	3.06	3.73	4.15	4.45	4.69	4.88	5.05	5.19	5.32	5.43
	.01	4.26	4.96	5.40	5.73	5.98	6.19	6.37	6.53	6.67	6.79
14	.05	3.03	3.70	4.11	4.41	4.64	4.83	4.99	5.13	5.25	5.36
	.01	4.21	4.89	5.32	5.63	5.88	6.08	6.26	6.41	6.54	6.66
15	.05	3.01	3.67	4.06	4.37	4.59	4.78	4.94	5.08	5.20	5.31
	.01	4.17	4.84	5.25	5.56	5.80	5.99	6.16	6.31	6.44	6.55
16	.05	3.00	3.65	4.05	4.33	4.56	4.74	4.90	5.03	5.15	5.26
	.01	4.13	4.79	5.19	5.49	5.72	5.92	6.08	6.22	6.35	6.46
17	.05	2.98	3.63	4.02	4.30	4.52	4.70	4.86	4.99	5.11	5.21
	.01	4.10	4.74	5.14	5.43	5.66	5.85	6.01	6.15	6.27	6.38
18	.05	2.97	3.61	4.00	4.28	4.49	4.67	4.82	4.96	5.07	5.17
	.01	4.07	4.70	5.09	5.38	5.60	5.79	5.94	6.08	6.20	6.31
19	.05	2.96	3.59	3.98	4.25	4.47	4.65	4.79	4.92	5.04	5.14
	.01	4.05	4.67	5.05	5.33	5.55	5.73	5.89	6.02	6.14	6.25
20	.05	2.95	3.58	3.96	4.23	4.45	4.62	4.77	4.90	5.01	5.11
	.01	4.02	4.64	5.02	5.29	5.51	5.69	5.84	5.97	6.09	6.19
24	.05	2.92	3.53	3.90	4.17	4.37	4.54	4.68	4.81	4.92	5.01
	.01	3.96	4.55	4.91	5.17	5.37	5.54	5.69	5.81	5.92	6.02
30	.05	2.89	3.49	3.85	4.10	4.30	4.46	4.60	4.72	4.82	4.92
	.01	3.89	4.45	4.80	5.05	5.24	5.40	5.54	5.65	5.76	5.85
40	.05	2.86	3.44	3.79	4.04	4.23	4.39	4.52	4.63	4.73	4.82
	.01	3.82	4.37	4.70	4.93	5.11	5.26	5.39	5.50	5.60	5.69

q Values

Percentage Points of the Studentized Range

Error df	A	\(r \) = number of means									
		2	3	4	5	6	7	8	9	10	11
60	.05	2.83	3.40	3.74	3.98	4.16	4.31	4.44	4.55	4.65	4.73
	.01	3.76	4.28	4.59	4.82	4.99	5.13	5.25	5.36	5.45	5.53
120	.05	2.80	3.36	3.68	3.92	4.10	4.24	4.36	4.47	4.56	4.64
	.01	3.70	4.20	4.50	4.71	4.87	5.01	5.12	5.21	5.30	5.37
∞	.05	2.77	3.31	3.63	3.86	4.03	4.17	4.29	4.39	4.47	4.55
	.01	3.64	4.12	4.40	4.60	4.76	4.88	4.99	5.08	5.16	5.23

Note: This table is abridged from Table 29 in *Biometrika Tables for Statisticians*, vol. 1, 2nd ed. New York: Cambridge, 1958. Edited by E. S. Pearson and H. O. Hartley. Reproduced with the kind permission of the editors and the trustees of *Biometrika*.

References

Abelson, R. P. (1995) *Statistics as principled argument*. Hillsdale, NJ: Lawrence Erlbaum.

Coolidge, F. L. (1983) WISC-R discrimination of learning-disabled and emotionally disturbed children: an intragroup and intergroup analysis, *Journal of Consulting and Clinical Psychology*, *51*, 320.

Coolidge, F. L. and Fish, C. E. (1983) Dreams of the dying, *Omega-Journal of Death & Dying*, *14*, 1–8.

Micceri, T. (1989) The unicorn, the normal curve, and other improbable creatures, *Psychological Bulletin*, *105*, 156–166.

Needleman, H. L., Riess, J. A., Tobin, M. J., Biesecker, G. E. and Greenhouse, J. B. (1996) Bone lead levels and delinquent behavior, *Journal of the American Medical Association*, *275*, 363–369.

Rosnow, R. L. and Rosenthal, R. (1989) Statistical procedures and the justification of knowledge in psychological science, *American Psychologist*, *44*, 1276–1284.

Cronbach, L. J. (1957) The two disciplines of scientific psychology, *American Psychologist*, *12*, 671–684.

Tukey, J. W. (1969) Analyzing data: Sanctification or detective work? *American Psychologist*, *24*, 83–91.

Tukey, J. W. (1977) *Exploratory data analysis*. Reading, MA: Addison-Wesley.

Tufte, E. R. (1983) *The visual display of quantitative information*. Cheshire, CT: Graphics Press.

Tufte, E. R. (1997) *Envisioning information*. Cheshire, CT: Graphics Press.

Index

design, 153
study, 142
interaction (effect), 205–206, 218
intermediate result, 259, 270

klinkers, 55
kurtosis, 44–45

language to avoid, 214–215
LDFA, 267, 271
least significant difference (LSD) test, 189, 194
leptokurtosis, 44–45, 49
likelihood ratio for a positive test, 259, 269
linear discriminant function analysis (LDFA),
    267, 271
linear relationships, 121–122, 135

main effect, 206, 217–218
MANCOVA, 265, 271
MANOVA, 264–265, 270–271
matched *t* test, 156, 170
mean, 51, 64
    hypothesis testing, 98
    unbiased estimator, 52, 64
measurement error, 16–17
measures of central tendency, 52–53, 64
    choosing, 54
    mean, 51, 64
        hypothesis testing, 98
        unbiased estimator, 52
    median, 52, 65
        method 1, 52
        method 2, 53
    mode, 54, 65
median, 52, 65
    method 1, 52
    method 2, 53
Mercury, Freddy, 262
Micceri, Theodore, 47
mode, 54
mortality and morbidity, 261, 263
    age specific death rates, 262
    annual crude death rate, 261–262
    case-fatality proportion, 263
    incidence rate, 262
    prevalence proportion, 263
multiple comparison tests, 179, 186, 189, 194
multiple regression, 266, 271
multiple *t*s, 172, 185
multivariate analysis of covariance (MANCOVA),
    265, 271
multivariate analysis of variance (MANOVA),
    264–265, 270–271
multivariate statistics, 256, 263–269
    a summary of, 268
    analysis of covariance (ANCOVA), 263–264,
        270
    canonical correlation, 267, 271
    cluster analysis, 267–268, 271
    factor analysis, 265–266, 271

linear discriminant function analysis (LDFA),
    267, 271
multiple regression, 266, 271
multivariate analysis of covariance
    (MANCOVA), 265, 271
multivariate analysis of variance (MANOVA),
    264–265, 270–271

Neyman, Jerzy, 105, 106
Nightingale, Florence, 23, 29–30
nondirectional alternative hypothesis, 99, 103–104
nonnormal distributions, 44–45
nonparametric statistics, 239, 252
nonsignificant differences, 193
nonsignificant findings, 102–103
normal curve, 37
null hypothesis, 95, 99, 106
    retain or fail to reject, 99–100
Nunnally, Jum, 47

omnibus hypothesis, 173, 185
omnibus test, 173, 185
operational definitions, 16
order of presentation, 170
outcomes, 101
outliers, 56

p level, 107
paired difference *t* test, 156
paired *t* test, 156
parameters, 24
    describing data, 63
Pearson, Egon, 91, 105, 106
Pearson, Karl, 23, 90–91, 105, 106, 134, 252
Pearson product moment correlation coefficient,
    115–120, 135
    significance, 118–120
percentile, 42, 43, 49
    summarizing scores, 89–90
phi correlation, 115, 132–134, 136
    significance, 133–134
placebo, 15
placebo effect, 8–9, 25
platykurtosis, 44–45
point-biserial correlation, 115, 130–132, 136
    significance, 131–132
pooled variance, 145
population, 24
post hoc tests, 179, 189, 194
power, 150, 153
    abuse of, 10
    analysis, 10, 150
prediction, 61
prevalence, 259, 269
prevalence proportion, 263, 270
probability, 101
    p level, 107

quartiles, 43, 49
Quetelet, Adolph, 23

random effects, 206
range, 57
receiver operating curve (ROC), 256, 269
regression effect, 156–157, 170
relative risk reduction, 261, 270
repeated measures design, 170
repeated measures *t* test, 156
replication, 100, 107
research hypothesis, 95, 106
risk assessment, 260–261
   absolute risk reduction, 260–261
   relative risk reduction, 261
robustness, 143
rounding error, 20
rounding numbers, 20

sample, 24
   representative of the population, 13
sampling distribution, 175, 185
scales
   interval, 19, 24
   nominal, 18, 24
   ordinal, 18, 24
   ratio, 19–20, 24
scatterplot, 120–122, 135
Schmidt, Johannes, 64
sensitivity, 257–258, 269
skewed distribution, 38
   negative 38–39
   positive, 38–39
signal-to-noise ratio, 95, 106
significance, 107
   critical value, 102
   nonsignificance, 102, 107
simple effects, 206, 218
Simon, Pierre, 23
Sinclair, John, 22
Snedecor, George W., 184
Snow, John, 31
Spearman, Charles, 134
Spearman's correlation, 115, 127–128, 136
   significance, 128
specificity, 258, 269
standard deviation, 57, 59
   computational formula, 60
   correction for bias, 59
   empirical rule, 61–63
   prediction, 61
standard error of the mean, 175, 186
standard scores, 69, 91
   T scores, 69, 73–74, 91
   z scores, 70, 91
      calculating, 71–73
      converting, 73–74
      formula, 71
      hypothesis testing, 76–77
      interpreting, 74–76
      negative, 75–76

summarizing scores, 89–90
statistical control, 14
statistical symbols, 20–22
statistics
   descriptive 5–6, 24
   inferential, 5–6, 24, 94
   purposes, 1–2
statistician, 2–4
   liberal and conservative statisticians, 4–5
stem-and-leaf, 43–44
subject error, 174, 185

tables
   purpose, 29
   summary of purposes, 32–34
T scores, 69, 73–74, 91
t test, 152
   dependent, 156–169
   independent, 141–150, 152
test of the standardized residuals (in chi square
     test), 250–251, 253
theory, 6
   oddball theories, 6–7
ties in ranks, 128–129
trend, 103, 107
true negative, 257–258, 269
true positive, 257–258, 269
Tufte, Edward, 29, 31, 32, 34
Tukey's HSD test, 189–192, 194
   with unequal Ns, 193
Type I error, 101, 103, 104, 107
   p level, 107
Type II error, 102, 103, 104, 107

uninterpretable result, 260, 270

variation, measures of, 57–61, 65
   range, 57, 65
   standard deviation, 57, 59, 65
   variance, 57, 61, 65
variance, 57

Walker, Helen, 91, 103
within-group error, 141
within-group(s) variance, 141, 174, 185
within-subjects *t* test, 156

z distribution, 60, 74–76, 91
   empirical rule, 75–76
   hypothesis testing, 76–77, 96
z scores, 60–90, 91
   calculating, 71–73
   converting, 73–74
   formula, 71
   hypothesis testing, 76–77
   interpreting, 74–76
   negative, 75–76
   summarizing scores, 89–90